The Mogollon Monster, Arizona's Bigfoot

Susan Farnsworth

ISBN: 1461016266
ISBN-13: 9781461016267

DEDICATION:

THIS BOOK IS DEDICATED TO MITCH WAITE.
WITHOUT HIS INSPIRATION AND ENCOURAGEMENT,
MY EXISTANCE, AS AN AUTHOR AND WRITER WOULD
NOT HAVE COME ABOUT. MY SINCERE GRATITUDE.

CONTENTS

INTRODUCTION

The Mogollon Rim is located in northern Arizona. It extends from the Flagstaff area eastward through Payson and into the White Mountains near the towns of Springerville and Arizona. The rim can also be traced into New Mexico.

The Rim is a part of the Mogollon Plateau, which was formed from major geological faulting in Arizona. It received its name from the Governor of New Mexico, Juan Ignacio Flores Mogollon, between the years of 1712 and 1715. The Rim crosses some of the most rugged terrain of Arizona much of which very rarely visited by man. It is in these areas that the monster is said to exist. From time to time, the monster has emerged from its hiding places to make an appearance. When it does, it leaves behind some bazaar tales by those who were there.

Many unbelievers state that the monster is a hoax. In those cases where people have been mauled or even killed, the authorities write it off as either a rogue bear, or even a crazed recluse or hermit. However, the incidents keep happening.

I was born and raised in Saint Johns, Arizona. While I growing up I kept hearing stories from the local towns people of a monster living along the Mogollon Rim. I had always assumed these stories to be those you would tell around the campfire until I had a spooky experience on the South Fork of the Little Colorado River.

I started collecting the stories for research. I wanted to compare the events to see if there were any similarities. The results were astounding.

Is it man or beast? What does it look like? Why have we not captured it? These are typical questions asked by believers and non-believers alike. The next question is how old is it? Since the sightings have occurred throughout history it can't be the same animal. Or could it?

What does the beast look like? That's an interesting question. In most incidents, the creature is said to be very hairy and walks upon its hind legs like a man. However, it's speed and strength eliminates the possibility of it being human. It is also described as being black or dark brown with no hair on the chest or upper face. It's hands and feet are also not covered in hair. Its eyes are said to be wild and red. One account even states that it is not even mammalian. Instead, it is reptilian in nature, and has no hair at all.

How old is it? Who knows. If it is a single beast, it would have to be hundreds of years old. If there are many, then the age could vary quite drastically. I would have to shun away from the very old theory, because most stories hint to the possibility of there being more than one.

Could there be more than one? Yes, there are sightings which have happened all over the US and Canada. There are the rock throwers near Modesto, California; Oregon has it's quiet and shy big foot; and Green Springs, Colorado has the garbage can raiders.

Arizona's big foot seems to display the same basic characteristics as the others with a few exceptions. It seems to be very territorial. It has both vegetarian and carnivorous diets. And, it is sometimes very violent.

So, is the Mogollon Monster real? You will have to decide.

SUSAN FARNSWORTH

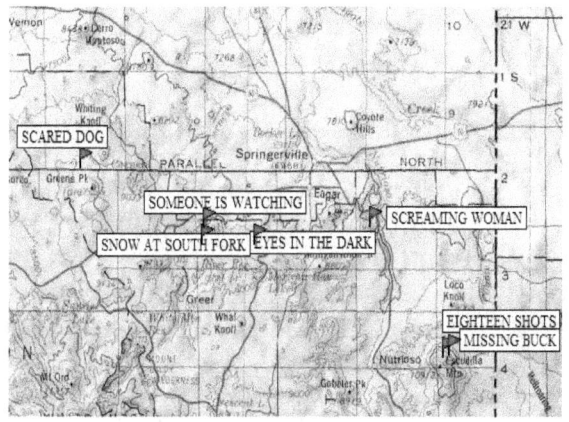

MAP 1

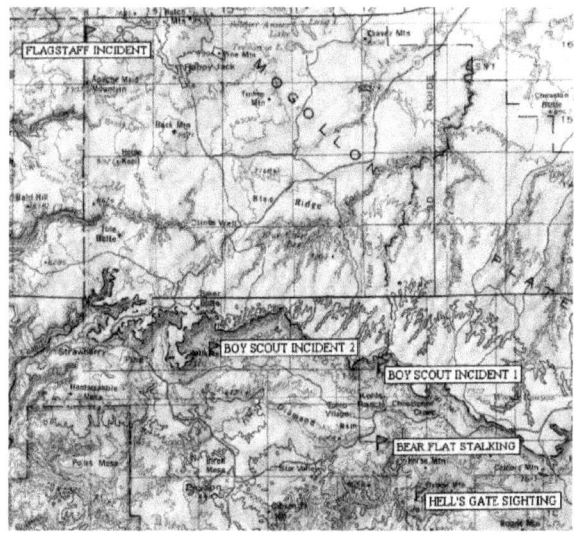

MAP 2

CHAPTER 1: EIGHTEEN SHOTS

THIS STORY COMES FROM JERRY WAITE. IT IS AN ACTUAL EXPERIENCE WHICH HAPPENED TO HIS FATHER, JAKE, WHILE HUNTING THE FOOTHILLS AROUND ESCUDILLA MOUNTAIN. JAKE WAS A FUR TRAPPER AND HUNTER BY TRADE. HE MADE HIS LIVING OFF THE LAND, AND KNEW IT BETTER THAN MOST PEOPLE AROUND. JAKE HAD HUNTED AND TRAPPED ALMOST EVERY TYPE OF ANIMAL THERE WAS IN ARIZONA UNTIL THIS DAY. THE DATE WAS NOVEMBER, 1923.

"I'll be back about dark tomorrow night," Jake called to his wife as he was leaving. "I've left the shotgun on the mantle."

Jake always left his loaded shotgun on the fireplace mantle since an ugly incident several years before. His wife and children had been threatened by an old, crazy hermit, but his wife insisted it was a monster of some sort. At any rate, the dog had saved the family and nothing was ever heard of from any person about anyone being chewed up.

However, Jake had seen and heard some curious things in the woods around Escudilla. Just to play it safe the shotgun would remain loaded on the mantle.

Jake walked out to his saddled horse and loaded the last of his supplies into the saddlebags. He slid his new thirty-thirty into the rifle scabbard and turned around to see his wife coming out the door with an extra blanket.

"You might get cold," she stated as she rolled the blanket into a tight tube. "You better take this."

Jake took the blanket from Lillian and tied it to the back of the saddle with his other blankets. "This time I'll get that ole bear." He smiled.

Lillian smiled. "Perhaps you shouldn't."

"Why?"

"Because every year the cattlemen have raised the bounty. He's worth what now...two hundred dollars?"

"That's right," Jake agreed. "But we sure could use the money now. The way prices keep going up, it won't be long before we'll be paying two bits for a loaf of bread."

"You know that won't happen," Lillian laughed. "Not as long as women can bake bread. That store bought stuff will never catch on."

Jerry stepped closer to his wife and pulled her to his side. They kissed for a moment. Jerry looked into his lovely wife's eyes and he wondered how he had been so lucky to land a woman such as this one.

"Well," He sighed. "I guess I'd better get going."

He turned to his horse and slipped into the saddle and rode away.

By nine o'clock Jake had made it past town and was topping Picnic hill. He had gotten a good jump on the day. He would be making the foothills of Escudilla by noon at this rate. He was hoping to be able to get some hunting done while the trail was still hot. One of the cattle ranchers had spotted a fresh kill near a prominent rock point on the mountain. It was Club Foot all right. The trail leading away from the mauled carcass had three normal bear prints and one stub.

"I'll pick up Club Foot's trail tonight, and get him tomorrow." He thought. "If all goes well, I'll be back tomorrow night and I'll be rid of that bear once and for all."

Jake had been hunting the bear for several years. It was believed to be one of the few grizzly bears spotted in Arizona. It had gained a taste for cow and had been killing quite a few of the live stock around the area. But lately, a peculiar thing had happened.

3

The bear had stopped eating the meat. It would kill the cows and mangle the carcasses. Very strange in deed. Jake had always assumed that it was because the bear had turned mean when it lost its foot. Since it was a man's trap (his as a matter of fact) that took the bear's foot, Jake thought the bear was getting even by killing the cattle that man was trying to grow. Maybe it had something to do with the weather. It was hot and dry. The maulings seemed to increase on hot and dry years. Perhaps, the heat was activating the bear's need to kill for meat?

Jake rode on. By four o'clock, he had reached the prominent rock outcropping. It didn't take long for him to locate the decomposing carcass under some low hanging trees.

He had found it by following his nose. The stench was surprisingly bad. To his surprise this wasn't the first animal to die beneath the branches of these trees. There was evidence of several other animals. One looked like a deer, maybe two elk, and several other cows.

"This must be the bear's favorite eating place," he spoke out loud to his horse when he climbed back into the saddle. "I've never seen anything like this," he mumbled. "Bears don't carry their kill. They usually bury it at the kill site and come back for it later."

Jake was a bit shaken by this turn of events. This was no ordinary bear. But then again, there weren't very many grizzlies in Arizona either. Was it something characteristic that grizzlies did?

Suddenly the horse stopped dead in its tracks. Its ears were perked to the front and every muscle seemed to tense up.

"What is it boy?" Jake whispered as he leaned forward to pat the horse for reassurance. "You sense something?"

Jake had long ago learned to trust the instincts of the animals around him. He knew they could sense danger well before man could. It was this fact that had helped him to become the successful hunter that he was. He had lived with his animals and knew them. He even felt sometimes he could read their thoughts, and this time it was no different. The horse was frightened. It definitely knew something was about.

Jake started getting the feeling too. Something was watching them closely. They were being stalked. Jake drew his thirty-thirty and shucked a shell into the chamber. He let the horse have its head to determine from which direction the danger was coming. Slowly the horse's ears began to move. The horse would turn keeping it's head pointing in the direction of the ears.

"That's right," Jake whispered. "You point him out to me. I'll do the rest."

The horse continued to move in a standing circle. The stalker seemed to be circling. It was looking for the moment and the weakness to attack.

Jake realized what was happening and knew the attack would come if the stalker was allowed to work its way up hill from them. It would have the cover of a thicket to get closer, and the attack would be faster since it would be down hill.

"Trust me on this one," he whispered to his horse.

Violently he jerked the reins to the left causing the horse to bolt downhill. Both horse and man were flying at breakneck speed down the hill side towards the outcropping of rocks. If they could reach the rocks there would be no way the bear could attack without exposing itself. This would allow Jake to get several clear shots at the animal.

Upon reaching the rocks, Jake slipped from the saddle and hit the ground running. It was almost a fluid motion and the horse darted below the rock bluff to hide. Once in the rocks, Jake made ready for the attack that was sure to follow. He pulled a box of shells from his jacket and set them on the rocks in front of him. The box contained the remainder of rounds most of which had already been placed in the rifle.

Jake waited. He knew the animal was out there somewhere. Why had it not pursued him? Most bears would have, but this bear was different. It seemed to know what to expect from humans. And, when the attack did not happen, Jake began to worry about who was hunting who.

The sun was getting lower in the sky. It had been an hour or so since the original encounter. Jake watched the shadows grow longer, and realized he may have to spend the night perched in the rocks. Perhaps, that was what the bear was waiting for?

A noise snapped his attention to a sharp focus. It was a rock sliding down the hillside from below. Jake turned quickly in time to see a dark figure slip behind a large boulder below. It didn't seem to move like a bear. When it ducked behind the boulder, Jake could have swore it had been on its hind feet. Could it be a man? No, of course not. The figure was too big for a man. It was more the size of a grizzly. But he could swear the figure he had glimpsed was more slender than a bear. Whatever it was, it was after the horse below him.

Jake called out to the figure. "If you are a man, speak up!"

There was no reply.

"I am armed, and I will shoot!"

Only silence.

"The horse won't do you no good. It's too afraid."

Still no reply.

"I'd hate to kill you over a horse!" Jake shouted.

At that moment, the figure rushed from the boulder. It was surprisingly fast and had already traveled half the distance to the horse before Jake could raise his rifle to fire. The first shot rang out and the figure fell. Instantly, it was back on its feet moving toward the horse.

A second shot was fired, and it seemed to have no effect. Jake shucked a third shell into the chamber and fired. It too seemed to have no effect.

Jake was sure of his shots. He knew he had to be hitting the beast square in the chest, but it seemed to be bullet proof. Jake emptied his rifle at the beast.

Quickly Jake grabbed the box of ammo and reloaded. It would be a race to see if he could fire again before the beast got to the horse. Jake reloaded the magazine and pulled up to fire. The figure was only a few feet away from the horse when the shot rang out.

A patch of red appeared on the beasts shoulder, and it immediately screamed. It was a terrible scream. It sounded much like a woman in distress, but it definitely had tones of pain mixed in with the cry.

Jake shucked another shell into the chamber and pulled down for another shot. But the beast was retreating and no longer moved over the ground in a smooth motion.

Jake fired again and again emptying his rifle a second time. But the beast never fell.

Grabbing the box of ammo again, Jake found two shells left. He decided not to fire them, but to save them in case of another attack.

The animal, if it was that, slipped into the trees and out of sight, and all became quiet.

"That was no bear," Jake mumbled to himself. "Twasn't man either."

Jake climbed out of the rocks and found his horse. He climbed aboard they made a mad dash back to civilization.

The trip to Alpine was uneventful. But, Jake couldn't help but wonder why his shots did not take the creature down.

He stopped near a construction site and borrowed three two-by-four stubs. After stacking them against a tree he fired his two last shots into them. Both bullets barely penetrated the first two-by-four. The powder in the shells had gone bad.

CHAPTER 2: BURN THE MOUNTAIN

THIS STORY IS A CONTINUATION OF 18 SHOTS. IT IS ONE VERSION AS TO WHY ESCUDILLA MOUNTAIN WAS BURNED IN THE EARLY 1900's. THE LOCALS WHO REMEMBER THE MOUNTAIN BURNING SAID IT COULD BE SEEN ALL THE WAY FROM SAINT JOHNS. WAS IT AN ACCIDENT? NATURE? OR AS THE STORY IS TOLD.

Knowing why the beast did not fall, Jake purchased another box of ammo. His next stop was to an old friend of his who was also in the hunting business, Elias Morgan.

Elias was a hard man. He didn't like people too much, but he was an excellent tracker, and he was the only one in the area who had trained hunting dogs. Jake found Elias at his trading post near Alpine just before dark.

"Elias," Jake called when he entered the door. "You ole son of a gun. How have you been."

Elias turned his head long enough to see who was approaching, but he went back to inspecting some new traps hanging on the wall.

"I've go a job for you," Jake stated.

"That's more like it," he mumbled and turned toward Jake with a smile. "What's in it for me?"

"Fifty dollars."

"Fifty dollars? Wha'd I haft ta do?"

"We got to track down a beast."

"Beast?"

Jake looked around the store to see if anyone was close by or listening. "Yes. I was hunting Club Foot over on Escudilla..."

"Club Foot ain't on Escudilla," Elias interrupted.

"What?"

"The bear was last seen over on Baldy near Maverick raiding the town dump."

Jake looked a bit embarrassed.

"You're after the monster ain't you."

"Yes"

"One hundred."

"Can your dogs track him?"

"Sure. They'll track anything. You got a scent?"

"Yes. I shot it this afternoon."

"You seen it? Most have only heard it or smelt it. Shot it, huh."

"Yes. I shot at it eighteen times, but the powder was bad. Only one, maybe two hit."

"That's bad. Maybe your shootin's gone bad. One hundred fifty dollars. Any animal's dangerous once it's been wounded."

Jake was a bit angry at this point. He was being fleeced, and he didn't like it. "I don't have that kind of money, and you know that."

"Well then get it."

"Better yet, I'll just go hunt it down by myself." With that Jake headed for the door.

"You'll need dogs," Elias stated in an effort to renegotiate. "I could go up there and get the animal myself."

"Go ahead," Jake growled. "It's a big mountain. Just don't get in my way."

Jake shoved the door open and stepped outside. He looked up the road toward Escudilla. He was contemplating going back for the animal, but really didn't like the idea of going alone. He needed help.

The door swung open and Elias stepped outside to join him.

"Partners."

Jake smiled. "Ok, then partners. We'll start out in the morning.":

Elias put his hand on Jake's shoulder and spoke, "You're the only man who gave me a square deal. You're the only one I'd trust to go after the monster."

"Why's that?"

"Because, you won't run. No man would go back into the mountains after the beast knowing it actually exists."

"What about the locals?" Jake asked. "They must know what it is."

"The Injuns know. They say to stay away from this area. They even put up warning signs on the rocks."

Jake looked at Elias curiously. "You mean the rock carvings down on the Little Colorado near South Fork?"

"That's right. They picture it with horns and claw like hands."

"Yes, I've seen it." Jake admitted.

"The Injuns call it the devil beast. They say it smells too."

"That it does," Jake laughed. "It's a foul smell of death."

"Some of the cattle boys have been talkin' bout torching the mountain to get rid of it. That is the ones who will admit it isn't Club Foot. That poor bear's been getting' a lot a blame for something it ain't doin'."

"Well," Jake sighed. "We'll just have to clear its name won't we."

Elias laughed. "That's what I likes about you Jake. You got good humor."

Jake smiled. "Say, partner. Do you have a place to stay the night?"

"Why sure. Just bring your blankets in here and you can sleep in the back room.

There's some sacks of beans that'll make a good bed."

"Thanks." Jake walked over to his horse and pulled off the blankets.

"Just pull the door shut when you come in," Elias commented as he headed back into the store. "Blow out the lamp when you're ready."

Jake took his horse around back and settled it into the corral. A few moments later he was fast asleep on the sacks of beans in the store.

The next morning came early. Elias was up and cooking some bacon on a pot bellied stove along with some coffee well before sun up. Jake got up and went outside to saddle his horse. While he was outside he noticed that Elias had already been there and done the chore. In fact there were several horses readied for the trip.

Jake stopped off at the watering trough and stripped to his waist. He splashed some water on his face and arms. The cold water felt good, and it helped get the circulation going. He dried his arms and face with his shirt and then put it back on. He was ready for breakfast.

"Grubs ready," Elias stated when Jake walked back in.

The bacon, biscuits and gravy sure hit the spot. Jake hadn't eaten since he left his home the morning before. Elias sure knew how to take care of a partner.

Within a few hours they were well on the trail back to Escudilla Mountain. Jake led the pack with Elias and his dogs in the rear. It seemed pretty much like any other day of hunting.

It was mid morning when they reached the rocks. Jake pointed out the area where the animal had been wounded. Sure enough there was blood on the rocks. Within moments of showing the blood to the hounds, they were off and running. The dogs seemed to be making a beeline up the slope to the top of the mountain. Jake and Elias where right behind them on their horses.

Once at the top of the mountain, the dogs seemed to pause to get their bearings, but were soon off and running again.

"They got a hot trail," Elias shouted above the noise of the rushing horses. "It won't be long now."

The hounds had changed their normal yapping to long howls.

Jake reached down and touched his rifle making sure it was ready. He noticed Elias had done the same. Jake noticed Elias' rifle. It was a Springfield 45/90. It would certainly do a job on any animal in these woods.

The riders came to s steep dip in the mountaintop and there were the dogs. They had surrounded a water hole. The ice on the top had been broken and it looked as if there was blood on some of the ice chunks.

"Well, I'll be," Jake spoke astounded. "This here is one smart critter we are following. It seems to know the ice water would stop the bleeding."

"How'd you know that?" Elias questioned.

"It's a trick a wounded bear would use. Stops the bleeding alright, and it confuses
the scent too."

Jake climbed from his horse and circled the water hole on foot looking for tracks. There seemed to be none until he found what he was looking for, the footprints of the animal coming into the water.

Jake shook his head in disbelief. "I don't believe it."

Elias rode his horse closer to Jake to see what he was looking at.

The footprints were almost human. Jake stuck his foot down inside one of the tracks, and it was much bigger than Jake's big feet even with the boots on.

"You sure you didn't shoot someone?" Elias teased.

Jake kept shaking his head in the negative mostly from disbelief. Then he walked up the back-trail looking very closely at the prints and finally stopping nearly twenty-five yards away from the water hole.

"It backtracked it's trail," Jake mused. "Here's where it split off the trail heading almost the way it came."

"You mean we must have rode right past it?"

"More than likely."

"It stayed down wind too. It must be aware of its smell." Elias commented. "Otherwise, the dogs would have caught it."

Jake agreed.

Elias called the dogs over to the new trail and they were soon to pick it up. Jake jogged back to his horse and climbed aboard to join the chase.

A half a mile down the trail the hunting party came to an abrupt stop. They were overlooking a large pine thicket. It was very dense and going would have to be on foot.

Both riders dismounted and pulled their rifles. The dogs moved in ahead of the hunters and were soon swallowed up by the thicket.

"I think we should hang high and wait for the dogs to flush it out," Jake whispered.

"You wait here, and I'll take those rocks over there. We can cover it all."

Elias agreed.

Jake quickly took his position forty yards away in the rocks. It was none too soon. The dogs had found the animal, and a ferocious battle was being fought in the trees out of the sight of the gunmen. The small trees were being torn apart in the fight, but neither hunter could see the progress. One could only judge the battle by the terrible growls and yelps from the fighting animals.

A blunt thud sounded and one of the dogs was silent, then another. This time the dog yelped in extreme pain. A few moments later the last dog was silent.

"My dogs!" Elias shouted in anger. "The monster killed my dogs!"

Elias stood up and fired into the thicket.

A tremendous scream rumbled forth from the thicket echoing down the hillside.

Elias started forward fumbling with his rifle to reload it. "Elias, stop!" Jake shouted.

But, it was no good. Elias stumbled closer to the thicket.

Jake started forward to try to intercept Elias.

"Don't go in there!" Jake shouted at the top of his lungs.

Within moments, Elias had disappeared into the thicket shouting profanities at the beast. Then a shot rang out followed by a cracking noise like a logger felling a tree. Elias' profanities stopped. Only silence followed.

Jake swallowed hard. What should he do? Go into the thicket to try and save his partner? Or wait and hope for the best?

Elias' words from the night before echoed through Jake's mind. "...You won't run."

Jake charged the thicket blasting a path with his thirty-thirty. This time he knew his ammo was good because he could see the wood splinters flying from the trees before him. Moments later he broke into a cleared spot where the small trees had been torn down.

Three dead dogs lay sprawled before him. Their bodies had been torn limb-from-limb. There was blood all over the ground, but no Elias. Jake didn't like this at all.

Picking a large pine tree to guard his back, he grabbed some shells from his shirt and reloaded his magazine. Meanwhile he searched the area of any sign of man or beast.

Jake spotted Elias rifle. It was broken in half. The stock was snapped in half. That explained the sound of wood breaking, but what happened to Elias? There was no sign of him. And, what of the beast?

Jake's heart was pounding hard in his chest. He thought for sure his heart could be heard for several yards. His senses were on full alert. He caught a light whiff of scent of the animal. It was up wind, but had to be some distance away. How did the beast do it? There were footprints all around the battle scene, but none leading away. Surely there would be drag marks if it were dragging Elias away.

Jake darted from the cover of the tree to a large fallen log. While working his way down the log he watched for any sign of movement. He spotted another large pine tree, which would offer him some cover. He moved quickly to it. With each change in position, Jake could see more ground through the sapling pines.

Jake dashed from the tree to another. From this position, and now he could see some clothing. It was Elias. It appeared as if he had been stuffed under a large log. His first impulse was to run to Elias to see if he was alive, but he knew it could be a trap.

Quietly, he worked his way around the area to the opposite side of the log. There seemed to be no sign of the monster. It had gone. Quickly he moved down the log and knelt near Elias' body. He started to turn Elias over to look at his face. Suddenly, Elias came to life and turned over and took a swing at Jake with a large stick. Jake ducked the stick and grabbed Elias.

"It's OK," He spoke softly to Elias.

Jake pulled him up and propped him against the log. He started to inspect Elias for damage, and was surprised at what he saw. There were no injuries, but Elias' hair had turned white.

Elias was trying to speak, "Ba..Burn." He stuttered "Burn the mountain."

Jake looked at Elias inquisitively.

"Burn the mountain," Elias blurted it out. "Burn the mountain, it's the only way to get rid of them."

"It's Ok," Jake insisted. "We'll get an army of hunters up here and we'll get the monster."

"No." Elias grabbed Jake's shirt. "No!" Elias went unconscious.

Jake grabbed up Elias and slung him over his shoulder. As quickly as possible, he backtracked through the thicket and up the hill to the horses.

Several hours later they arrived at the trading post. Elias was still out, but alive. Elias never told what had happened in the thicket. No one knows how he got to the log or why he was stuffed under it. Some say is the way with bears to bury its prey to come back to eat it later. We will never know because Elias never talked again.

But some of Elias' last words stuck in Jake's head. "...to get rid of them."

CHAPTER 3: SOMEONE IS WATCHING

THIS STORY CAME FROM FRED ALBERT WAITE. IT WAS AN ACTUAL OCCURRENCE THAT HAPPENED TO HIM WHEN HE WAS A YOUNG BOY MUCH TOO YOUNG TO REMEMBER WHAT HAPPENED. IT WAS HIS MOTHER WHO RELATED THE STORY TO HIM YEARS LATER WHEN HE WAS OLD ENOUGH. THE DATE WAS AROUND SPRING OF 1926. THE LOCATION WAS THE SOUTH FORK OF THE LITTLE COLORADO RIVER EAST OF SPRINGERVILLE.

It was a beautiful spring day. A perfect day to let the boys out to play in the meadow while she swept the front porch of her new cabin. From the front porch Lillian could see most of the meadow to the north and east. It was a pretty view overlooking the Little Colorado River. Around in the back the meadow stretched for another hundred yards. It was perfect for watching the deer and elk grazing early in the morning.

Lillian was fairly new to the area. Her home had been over in Saint Johns thirty miles to the east, until she married her husband. They had decided to make a fresh start near the mountains where the hunting would be good. After all, that's what her husband did for a living. Jake was a fur trapper and hunter.

With the marriage, she gained some new responsibilities, Billy and Fred. They were the typical four-year-old boys always looking for something to get into. But, she loved them as if they were her own.

It was going to be a lovely cabin. It would have two rooms and an upper loft for the boys to sleep. For the last few nights, they had to sleep under the stars because the roof had not been installed. But, the weather had been good to them. In fact, it had been an extremely dry winter, and the spring showed no signs of change. Within a few days the log walls would be completed with the mud chinks between the logs and the roof would be in place.

"Then it could rain." She thought to herself.

It was unfortunate that her husband, Jake, had to go hunt that ornery old bear, Club Foot. But, that was his trade. He was the best trapper in the parts, and he had been hired to hunt down the bear because it had been mauling cattle.

The cattlemen in the area had put a one hundred-dollar bounty on the bear. If anyone could collect it, it would be Jake. The hide would bring a nice price too. Maybe enough to finish the cabin. At any rate, Jake would be coming home soon because he had been gone for over a week.

The boys had been playing in the meadow for several hours near the tree line to the north. Now she noticed they were heading back to the cabin in a hurry. "They must be getting hungry," she thought to herself.

As the boys got nearer they began to call out to Lillian.

"Mom!" Billy called frantically. "There's someone watching us."

Fred was right behind Billy running as fast as he could. He too was jabbering something about a man in the trees.

"Well," She shouted. "Who is it?"

"Don't know," Fred answered.

Jerry interrupted, "It's a big harry man, Mom! and he ain't go no clothes on."

Jerry's answer troubled Lillian. She didn't know everyone here in these local parts, but she was trying to place a big harry man in her mind when she spotted a rather large, dark figure paralleling the tree line coming towards the cabin.

The figure was upright, not on all fours like a bear, more like a man. Was it a man? She could not tell, but it was very hairy looking and very big.

"Hurry boys" She grabbed her broom and started towards them as if to make them run faster. "Get in the house!"

Lillian strained her eyes trying to get a better look at the figure, which was now coming up a shallow ravine towards the cabin. It had no clothes. Not even a loincloth. It seemed awkward, but its speed was astounding. It moved over the ground at near the speed of a horse.

She finally reached the boys and passed by them. She could tell that the creature had fixed its attention on the boys and didn't see her coming. She gave her broom a mighty swing, which totally caught the beast off guard and knocked it to the ground.

In one fluid motion, Lillian had changed directions and was pursuing the boys towards the cabin.

"Faster," she cried. "You must run faster."

Suddenly there was a hair-raising scream. Lillian could tell it had come from the beast, but she didn't slow down to take a closer look.

It almost sounded angry. It was now getting close enough, she could hear its hard sounding footsteps behind her.

It seemed like an eternity, but the boys were finally climbing the stairs to the porch of the cabin. She hit the steps grabbing the two boys and flung them into the house in front of her. She immediately slammed the door behind her.

There was another terrible scream. This time it seemed to have tones of frustration in it. She could hear the beast near the side of the house. It began circling the log cabin looking for a way in.

Every now and then she could see a pair of eyes peering at her through the cracks between the logs. The eyes were cold and brown and had a terrible anger in them. There was a awful stench in the air much like an animal had died. And, there was the horrible rasping noise it was making when it breathed.

Lillian found herself next to the fireplace. She felt much better when she found the old shotgun. At least if the animal were going to break in, she would make sure it ate plenty of buckshot prior to it getting her and the kids.

The animal worked its way around the house and on to the porch. It was working its way toward the door. This is where it would attempt to get in, it was the weakest spot in the cabin.

Then a familiar growling could be heard. It was that of her old mother dog, Lady. At first it was a low growl, but the closer the beast got to the door, the louder the growl became.

The fight came quickly. The dog attacked the beast tearing at its limbs. The beast backed up and howled at the pain. Then it counter attacked. It swung a cross blow catching the dog across the body and sent it flying off the porch. The dog yelped in pain.

The beast started for the door again, and the dog attacked again. This time the dog went for the throat trying desperately to catch the jugular vein. The beast managed to slip the dogs grasp and send it flying from the porch once again.

The beast moved toward the door, and began to try to push it open. Lillian grabbed a log bench and jammed it against the doorknob to keep it from opening. The door seemed to bulge from the weight of the monster.

Suddenly, the dog attacked again. This time it had managed to attack from behind. Its teeth sinking deeply into the monster's flesh near the neck. The beast howled in pain and tried to throw the dog to the dirt, but it was futile. The dog and locked its jaws on the monster and was hanging on like a leach.

The monster retreated from the porch, trying to reach behind it to pull the dog free with no success. Realizing the only way to rid himself of the dog was to retreat, the monster headed for the tree line.

Half way across the meadow, the dog let go and fell to the ground, but the monster had been defeated. It continued its rush towards the trees and quickly disappeared.

Lillian could see the poor dog lying in a heap in the grass, but she dared not risk coming out of the cabin. Finally, she could bear it no more. She grabbed the shotgun and headed for the door.

"You boys bar the door when I leave."

"Yes, ma." they replied.

"Don't you open it until you hear my voice."

Again the boys agreed.

Lillian started out for the dog. It seemed an eternity as she crossed the meadow to the pulp of flesh, which use to be her prized hunting dog. She carefully inspected the dog while keeping her eyes on the tree line. The dog was dying. It had several bones broken and one protruding from its side.

Lillian was amazed at how the dog had given its last breath of life defending her and her children. The she looked into the glazing eyes to see the last spark life drain. It was dead.

As she squatted near the dog, she saw movement in the tree line. Instantly, she drew her gun and pulled down on the approaching figure.

"Who's there!"

"Don't shoot. It's me," came a familiar voice. It was Jake

CHAPTER 4: THE SCREAMING WOMAN

DONNA BIGELOW TOLD ME THIS STORY. SHE GREW UP IN THE TOWN OF SPRINGERVILLE, ARIZONA. SHE TELLS OF A NIGHT THE TOWN WILL NEVER FORGET. THE EXACT DATE OF THE INCIDENT IS UNKNOWN, BUT IT WAS SAID TO HAVE HAPPENED IN THE EARLY SUMMER OF 1938.

"Good evening," the young lady greeted the cowboys entering the church. "You'll have to check your guns."

The two cowboys looked at each other and shrugged. Then looking at the young girl, they decided to check their guns.

"Are all the girls as pretty as you?" One of the young riders questioned.

The girl blushed. "Of course."

The cowboy grinned at the other while he finished stripping off his holster.

"You were right," the other cowboy stated. "This is a nice church. Mormons huh?"

"Yup," the other answered. "They have the dance every Saturday night, but they don't go past eleven o'clock."

"Why's that?"

"I don't know. Has something to do with the Sabbath."
They passed their pistols to the young girl and turned to face the inner part of the building.

"Well," the tall one sighed.

"I'm not a religious man, but I like a good dance."

The two walked into the church and were soon having a good time.

Several minutes later, a strange sound could be heard. It sounded like some type of washing machine, but the motor backfired too many times. Outside, an old Model T pulled up to the front of the church. The commotion seemed to frighten the horses tied at the hitching post, but they soon calmed down.

A handsome young man with coal black hair stepped from the car. He walked around to the opposite side and offered his assistance to his date still in the car.

"Ma-dam. The ball awaits you." He comically gestured. But, the look in his eyes was that of total love.

Donna's eyes met his. There was a long pause as she looked into those dark blue eyes. She wondered about what she saw in those eyes.

"The man behind those eyes," she thought.

He was the only man her age who owned an automobile. He was different. He liked to tinker and fix things. The others were intent to ride the range and tend to their farms. This man was a thinker. He built his car from parts of cars from scrap piles from several different nearby towns. He had built something from nothing.

Donna was a stunning young woman although she didn't seem to know it. Her beauty had gone unnoticed by most people until recently. She had made friends with this new gentleman who seemed to bring out the best in her. Tonight she was radiant, and it definitely showed. Many of the young men were seeing her with new eyes. She was no longer the skinny little girl from the Bigelow clan. Her red hair was tied down with a scarf, and she was trying to take it off when her date offered her his hand.

"Just a moment," she fussed. "I want to get this thing off before we go in."

He stepped closer to her and studied the knot in the scarf. Gently he reached up and pulled the knot loose.

"There."

"Thank you kind sir." She whispered as she pulled the scarf away from her hair.

Jerry smiled and took her hand helping her out of the car. The two strolled into the church arm in arm.

"Oh, look." Donna stated excitedly when she entered the church foyer. "Everyone is here."

Jerry looked the crowd over and pointed out two couples. "That's something you don't see every day"

"Oh?"

"The Cosbys and the Halls in the same room together. And, they seem to be enjoying each other."

"This is true," Donna agreed. "But, everyone puts their differences away when it comes to the dance."

"Speaking of dancing," Jerry gestured toward the dance floor. "Shall we?"

The two stepped out onto the floor and began to waltz.

Suddenly, the hat check girl burst onto the dance floor and was screaming. It took a few moments for the crowd's noise to die down to where here words could be heard.

"There's a woman screaming for help!" She cried frantically. "Someone help her."

The two cowboys stepped through the crowd and started questioning the young lady.

"Which direction is the cries comin' from Ma'am," asked the older.

The young hat check girl seemed a bit frustrated, but she started explaining over again. "I was standing at the front door when I could hear a woman scream."

"Were there any words?" Someone in the crowd questioned.

"No, just a terrible scream. It sounded as if they were killing her."

"They?" asked the young Cowboy.

"There must be a couple of them," answered the young girl. "Her screams seemed to be moving away from here. I think she was being drug away!"

The two cowboys rushed for their guns and began to strap them on. They were followed by several of the other men who had previously checked their guns at the door.

The younger cowboy began to bark orders as if he was in the army, "You men without guns. Make sure the women folk get home and bar their doors! All the gun-totters follow us."

"We'll try to head the kidnappers off," shouted the older cowboy. "You men without guns, get some torches and your rifles and catch up to us as quick as you can. We may need them if we don't catch them in time."

Donna looked at Jerry.

"What do they mean in time?"

Jerry whispered, "The kidnappers will probably kill the lady when they are done with her."

Donna gasped.

"They won't want any witnesses," he continued.

The crowd was now pouring out onto the front lawn of the church. People where splitting up heading for their horses and homes.

An eerie scream could be heard in the far distance. It sounded as if it was near the sawmill.

"Come on boys," the older cowboy shouted. "They're getting away!"

A make-shift posse mounted their horses and were off in a rush towards the sawmill east of town.

Jerry assisted Donna into the automobile and walked around to the driver's side. He was shaking his head in the negative.

"What's wrong?" Donna queried.

"I don't like it." Jerry replied in the most preoccupied mood.

"Neither do I," Donna stated. "It's a shame this kind of thing could happen in such a nice town."

Jerry snapped his attention back to reality. "No. I mean, yes it is too bad, but I think something else is wrong."

"Really?"

"Yes. I've heard that scream before."

Donna was a bit puzzled. "You have?"

"Yes. My dad was a fir trapper by trade, and he use to take me out with him to check the traps. I heard that banshee cry once when I was alone in the woods."

"Well if it isn't a woman, what is it?"

"Dad said it was the Monster"

"The what?"

"He the Indians call it the Hairy Man. It's a beast that walks on its hind legs like a man, but it isn't."

"Oh come on...You got to be pullin' my leg. Aren't you?"

"No, I'm serious. My dad said he seen it one time and shot it eighteen times with his thirty-thirty. Then another time it came to our house when I was just a little kid. Mom said it meant to drag one of us little kids off into the woods."

Donna's face began to show some concern. "Why would it come to town?"

"It's really dry this year," Jerry stated. "Whenever it gets really dry, the maulings begin."

"Maulings?"

"Yes," Jerry continued. "The dryer it gets, the more the maulings move away from Escudilla towards the plains. It might have something to do with the availability of water."

Donna looked perplexed. "Maulings?"

"Yes." At first the ranchers were noticing dear and other animals were being mauled by what they thought was bears. Then it started in on the cattle. That's when the ranchers started hunting any bears large enough to maul a cow."

"But it wasn't a bear?" Donna asked still confused.

"No. Many of the people here thought my dad was crazy when he told them it was the Monster."

"Later, my mom had a confrontation with a huge man like creature which was trying to break into our cabin. She could see its eyes peering at her through the chinks in the logs of the cabin. She thought dad was crazy until that day. Anyway, the ranchers banned together and tracked the animal to Escudilla Mountain and they decided to burn the mountain."

"I always wondered why the mountain was burned," Donna was talking to herself. Then she grabbed Jerry.

"They didn't get the monster did they?"

Jerry shook his head in the negative.

She again reasoned things out loud, "It is hot this year. Too hot, and too dry. It's come out of hiding to look for food."

Jerry and Donna looked at each other surprised.

"We've got to do something," came her conclusion. "The posse is being lured into the mountains for an ambush!"

The two climbed into the automobile and started for the sawmill.

"Is this as fast as it will go?" Donna shouted to Jerry over the roar of the wind and engine.

"This is it," Jerry replied. We're pushing thirty five miles an hour! Any faster and the engine will blow!"

The two daredevils rounded the last corner and headed up the hill toward the mill.

"There they are!" Donna shouted while she pointed out the last of the posse.

"They've stopped to tighten their cinches."

Jerry started hitting the horn to catch their attention. But, the posse started to mount their horses again. Finally, one of the cowboys heard the commotion and swung his horse around to look at the oncoming rush. He signaled the others to hold their horses prior to going into the woods.

The automobile came to a sliding halt yards away from the posse.

"Don't go into those woods!" Jerry commanded. "It's a trap!"

"You've got to be kidding," the older cowboy laughed nervously. "The kidnappers are halfway up Picnic Hill. We'll never catch them if we don't hurry."

"No. I'm not kidding. You're being lead into the thicket."

"What?" The cowboy asked in unbelief.

Jerry knew he was going to be ridiculed for bringing up the Mogollon Monster.

He swallowed hard and began his explanation. "Don't you think it's peculiar that you haven't caught up with some kidnappers who have been dragging a screaming woman along with them?"

"Well," the younger cowpoke broke into the conversation. "We're about to catch them."

"Don't you think they would have gagged her by now? After all, she keeps giving their location away."

The crowd began to look at each other as if they were searching for answers. A young farmer by the name of Fred climbed on to his horse. "Well, it sure sounds like a woman in trouble."

"That's just it," Donna replied. "You are being lured into the mountains and it's not a woman!"

"Well," Fred started almost clearing his throat. "If it ain't a woman, what is it?"

Jerry stepped into the crowd and began to explain. "Remember about ten years ago, Escudilla burned."

"You ain't going to tell us that it was that monster thing are ya," a voice came from behind Jerry. It was the Harold Crosby. He was the store clerk in Eagar's general store.

"My old man told me that was because some of the cow pokes around here decided to get rid of Club Foot. Shoot, he weren't no monster. Just a bear that chewed his foot off in one of your dad's steel traps."

"You're wrong." Jerry stated flatly. "Yes, it was one of my dad's traps that took the bear's foot, but Club Foot was killed by my dad the year before the mountain burned. Hishide is on my wall in Saint Johns. The cattle ranchers burned the mountain to get the Monster."

"Bull!" The older cowboy coldly stated. "A bunch of bull pucky. There ain't no such thing. The only thing not human that could scream like that is a mountain lion, and that ain't likely. I'm going after it, man or beast!"

"Don't go off half-cocked." Jerry cautioned. "I can prove it's something trying to get you all into the mountains."

The eyes and ears of the entire group were on Jerry.

"You've been following the screams, Right?"

The group agreed.

"The screams have always come when you have stopped, wondering what direction to go next. Right?"

"The group mumbled in general agreement.

"If it's someone dragging a woman up the mountain, the next scream should be further away near the top of the mountain. Wait for the next scream. If it's going away from us, we'll go after it."

"And if it isn't" questioned the younger cowboy.

"If I'm right, the next scream will be closer to us than the last. If it is, we leave for town immediately. We'll chase this phantom in the day."

Suddenly a blood-curdling scream sounded just a few yards beyond their sight. The entire group turned to see if they were being attacked. Within moments, the posse was beating a hasty retreat back to town.

ARTIST CONCEPT OF
MOGOLLON MONSTER

CHAPTER 5: EYES IN THE DARK

THIS STORY CAME TO ME FROM MARVIN RAY. HE GREW UP HUNTING AND TRAPPING WITH IS FATHER IN THE WHITE MOUNTAIN AREA. THIS EVENT HAPPENED AT BIGELOW SPRINGS JUEAS EAST OF SPRINGERVILLE ON THE HIGHWAY TO BIG LAKE. MARVIN WAS TEN YEARS OLD WHEN HIS DAD TOOK HIM TO CHECK THE TRAPS. THE DATE WAS 28 SEPTEMBER, 1938

"Let's go," Marvin pleaded with his dad. "I want to go too."

Marvin was a young enthusiastic lad who loved to go with his father to check the traps. He always thought it was like Christmas. You never knew what kind of surprise would be waiting in the trap until you checked them.

Marvin's father, Johnathan, was preparing to go check the new trap line he had set around Bigelow Springs. He was hoping to bring back a rabbit or two for dinner. He had set the small game traps in some rabbit runs as well as baiting some traps for the meat eaters. Some fox furs would help purchase another bag of beans for the upcoming winter.

"Ok," Johnathan insisted. "We'll leave in a few minutes. Have you got your gun?"

Marvin pointed out his new single shot twenty-two rifle leaning against the wall near the supplies they had packed. "I'm ready."

"Here's your bullets," Johnathan stated as he passed a box of twenty-two bullets to his son.

Marvin did own his own gun. In fact it was more or less a tradition to train each family member in the use of firearms for hunting purposes. Along with the gun came certain responsibilities and lots of hands-on training. However the bullets were usually held in a safe box until it was time to go hunting. Marvin had proven to be reliable and safe when handling his firearm. Therefore, it was time to start trusting him with more responsibilities. This time, the boy would be able to carry the ammo while in the pickup and while out hunting. Johnathan wanted to observe Marvin to see if he could determine the proper time to load and unload the weapon.

Marvin looked the box of ammo carefully and then put it in his coat pocket. He walked over to the twenty-two rifle and picked it up. He made sure the gun was pointed in a safe direction and then opened the bolt to make sure it was unloaded prior to carrying it to the vehicle. He had been strictly trained to check the gun every time he picked it up or it was handed to him.

"Too many people had been shot by empty guns." Ran through his mind.

The two hunters placed their gear into the truck and climbed in for their trip. It wouldn't take long. They were only five miles to the trap line as the crow flies, but they would have to circle around following the dirt trail crossing the Little Colorado River near the old irrigation diversion dam a few miles south of Eagar. The trail would wind up the mountains to the first available watering area of Bigelow Springs.

It was a perfect day for hunting. Just after crossing the creek, Marvin spotted a cottontail rabbit and was allowed to take a shot.

"You only get one shot," his father told him. "You got ta give the game a chance. When you shoot an animal, you got to kill it with the first shot. It don't need to be sufferin', and bullets are too expensive to waste."

"Yes, Pa"

Marvin took dead aim and dropped the rabbit with one shot.

"Good shootin'" His dad told him. "The rabbit didn't even feel it."

They walked over to the rabbit and inspected the kill. The bullet had passed right through the rabbit's head. It didn't suffer.

"Why do we hunt with single shots?" Marvin asked. "I've seen those fancy lever action rifles that can put out 18 shots in a matter of seconds."

His dad looked at him as if he didn't believe what he had heard. Then he explained. "Only city folk hunt with those kind of rifles cause they're bad shots. They need more bullets to get the job done. I've seen one shoot five rounds before he even hit the animal. Then it was a gut shot."

Marvin shook his head in disbelief.

"What are the three rules for killing an animal?" Johnathan questioned.

"When the meat is needed," Marvin recited. "When the kill is sure and painless. Or when your life is in danger."

"Good." Johnathan agreed. "And when should you not kill?"

"When the shot is not sure, no gut or butt shooting. For sport, nor pleasure."

"Very good," Johnathan smiled at his son. "If everyone would practice this, the game and the forests would last forever."

"But not everyone does," Marvin commented.

"That's right. That's why we have game wardens. They make sure the city folks play by the rules."

Marvin looked confused.

"What's wrong son?"

"Mark Crosby says that his father is a game warden."

"So?"

"He says that he has to catch the local poachers."

"That he should."

"He says that anyone without a license is a poacher."

"No, that's only for the city folk. We grew up here. We depend on the land for our living. Without hunting we would starve. Besides, no game warden in his right mind would come up here in the mountains and tell me I couldn't hunt."

"Mark's dad says that some day we will have hunt only during certain seasons. And in the future, we might have to apply for a chance to hunt."

"Oh?"

"He says that there will be too many hunters for the animals so they will have to have drawings to hunt."

Johnathan looked concerned now. "You know, Mark's dad might be right. But, I hope I'm long gone from this world when it happens."

The two retrieved the rabbit and went back to the truck without saying any words. Soon they were headed down the trail towards the springs.

"Dad." Marvin broke the silence.

"Yes, son."

"What do you think it will be like when we can't hunt when we want?"

"Well," Johnathan thought out loud. "Now, if you go camping and you can see your neighbor's campfire, you're too close. Then I imagine, you will have to camp in designated camp spots right next to your neighbor so they can cram as many people in as they can."

"Sounds like a backyard sleep-out to me." Marvin stated.

"And I suppose since there will be so many, someone will come up with the idea to charge a fee for camping. Soon the land will not belong to the people any more. It will belong to the state or maybe the Federal government. When that happens, I can see all kinds of problems."

"What kind of problems?"

"Whenever a few people have power over many, the few become corrupt. Before too long, the few start acting like it's their own personal property. They start excluding the many."

"So the more people there are," Marvin questioned. "We will go camping less?"

"Yes. But, I hope that will be many hundreds of years from now."

The two sat contemplating the future as they drove on towards the springs. Arriving at their destination, they set about to set up camp. Marvin's job was to collect the firewood. He had to obtain enough for the night and the next morning's cooking fuel.

Johnathan finished the tent and prepared to go check the traps. "I'll go check the traps while you finish."

"Ok," Marvin agreed. "I'll finish gathering the wood. When do you think you'll be back?"

"I should be back around dark. If I'm not, start the fire and I'll be comin' in on it."

"OK, pa. I'll have the rabbit ready."

Johnathan pulled on his pistol and departed the camp.

Marvin finished up the wood gathering and set about skinning the rabbit. It would make a good Dutch oven stew when mixed with a few carrots, potatoes, and onions. Marvin peeled the potatoes and chopped the onions and carrots. He then diced the rabbit up carefully removing all of the bones.

Afterwards, he walked to the spring to obtain a bucket of water. While he was there he instinctively checked the ground for prints. Marvin spotted the usual prints. There were a couple of deer, a raccoon and maybe a coyote. But there was one print in the mud that seemed out of place.

Someone had been swimming in the pond. Several bare footprints were clearly visible. Marvin thought it was kind of peculiar that a grown adult would go skinny-dipping in a pond. Usually that was something that was reserved for the kids. But, these prints were adult. In fact, the size of the print would mean the person was roughly twice as large as his father. A big man indeed.

Marvin walked back to the camp and noticed the sun was nearing the horizon. The sun would be going down in a half an hour. He had better get dinner going.

The rabbit smelled good searing in the Dutch oven. It mixed with the smell of the bacon used to keep the meat from sticking to the pot. When it was well browned, Marvin mixed in the potatoes, onions, and carrots along with a large dose of salt and pepper. He placed the top on the oven and shoveled coals on to the top to evenly cook the stew.

It was starting to get dark when the stew was finished. Marvin pulled the pot off the fire and set it aside letting the top coals to keep it warm. It sure smelled good, but there was no sign of his father.

A half an hour later he decided to eat his portion of the meal. It smelled too good to wait, and he filled his plate. After he finished, he washed his dishes in the water from the spring. He laid down in the back of the pickup and looked up at the stars. It was a beautiful night. Too bad his father was not there to lay back and enjoy it.

Suddenly he awoke.

"I must of dozed off." He thought. He looked around and found the camp was nearly dark. The fire was flickering its last bit of light.

He quickly got up and placed some wood on the fire to get it going. He knew his dad would be looking for the fire and using it as a guide to come back to the camp.

Faintly, Marvin could smell something peculiar. It smelled like someone who had not had a bath in a very long time. The odor was getting stronger. Whatever it was crossing up wind of camp.

Marvin remembered what his father had told him many times before. "There's not much to be afraid of in the woods. Most animals won't come into a camp which has a good campfire."

Still, he couldn't shake the feeling he was being watched. He tossed a couple of heavy logs on the fire to keep it going. Afterwards, he walked over to the truck and retrieved his rifle. He opened the breach and checked to see if it was loaded. It wasn't. He made sure the bullets where in his pocket where he could find them.

A twig snapped on the hillside up wind of the camp. Something was out there all right. Was it his father? Or maybe the person who had been swimming in the pond? Maybe it was a bear?

Marvin was getting spooked. But, he kept his nerve.

"It might be dad testing me to see what I'll do by myself." He thought.

Just in case, Marvin propped his gun against the truck and set the bullets next to it. He moved over to the fire and placed more wood on it. The ring of light grew around the camp and Marvin could see further out into the woods. He remembered not to look into the fire, but to keep his eyes facing the dark to keep them adjusted for sight.

Then came the funny noise. It was kind of a thud on the ground next to him. Someone was throwing rocks! Marvin quickly moved over to the truck and retrieved his rifle and opened the box of bullets.

"Whoever is doing that better stop!" Marvin shouted. "I'm armed! And, I'll protect myself!"

There was no reply.
"Dad. If that is you, you better quit."

There was no verbal reply. But another rock came flying into camp barely missing the campfire. Marvin climbed into the back of the truck and watched from the protection of the fenders.

Another rock came flying into the camp. This time it hit dead center in the campfire spewing sparks and coals all over the camp sight. The fire was nearly out.

Marvin was scared stiff, but he realized that the fire had been the target. Whoever it was wanted the fire out.

Marvin sprang from the back of the truck and started piling the wood on the fire. Soon he had a raging bonfire that no rocks would be able to extinguish. He quickly retreated to the back of the pickup to watch what would happen next.

As the fire grew larger, the circle of visibility grew. Soon Marvin could see quite a distance up the hillside. He thought he saw a shadowy figure move from behind a bush to a large tree. He watched the tree very closely. But, he soon realized that all the wood he had was now burning on the fire. After it was gone, there would be no more.

Marvin watched the tree very closely. He was wondering what would act in such a manner. He even thought it might have been his imagination. Even so, someone had been throwing rocks.

The fire burned brightly for nearly twenty minutes and it began to die. The circle of light began to retreat. Soon the tree could no longer be seen. It was then Marvin heard something move in that direction. Whatever it was, it was moving closer to the camp as the fire died down.

Marvin tried to follow the movement with his eyes and ears. And, soon he could see two eyes staring at him from the bushes across the campsite. It was only fifty feet from him. This was not a wolf or bear. It was tall. The eyes glowed at him from above the bushes on the opposite side of the fire.

Marvin took his first bullet and loaded his gun. He carefully propped several other bullets up to here he could quickly load them into the breach.

Again he searched for the eyes. This time he located them peering at him from a "Y" in a large tree. It was moving in closer.

Marvin called out, "Don't come any closer! I'll shoot!"

By this time Marvin was convinced that the creature was not his dad. Perhaps his father was dead? Marvin shuddered from the thought.

The fire was hardly going when the eyes moved again. It seemed to be circling the campsite moving towards the pickup. Marvin removed the safety, and prepared to shoot. He aimed as close as he could to the movement in the dark. The next time the eyes would appear he would put a bullet right between them.

Suddenly, a noise came from the opposite side of the camp. It caught Marvin by complete surprise. And, as he swung his barrel around to fire something grabbed his gun.

Marvin screamed.

"What in the blue-blazes are you doin'?" Came a familiar voice.

"Dad?"

"And who else would it be?" His father was standing next to the truck looking down and Marvin now hiding in the bottom.

"There's something out there," Marvin insisted. "It's moving in."

"There's nothing out there," Johnathan replied. "Whatever was, it is gone."

Marvin peered over the fender of the truck at the fire. "Dad, there's no more wood."

"I thought so," his dad laughed. "Got spooked did ya?"

"There was something out there," Marvin insisted. "I saw it's eyes looking at me from the bushes. I think it's the Monster."

"Probably a coyote," His dad laughed.

"No, it threw rocks at the fire."

"It did what?"

"It almost put out the fire by throwing rocks at it."

Johnathan was now quite perplexed. His son had never lied before, and something had scared him quite badly. Yet he had heard stories of the monster in these parts, but those were old wives tales. Johnathan retrieved an old newspaper from the back of the truck and stuck it on the fire.

"There," he said confidently. "That will give us enough light for us to see to gather some more wood. You stay closely to the camp. I'll get the wood."

By this time the fire had caught the paper and began to burn more brightly.

"Dad."

"What?"

"What ever it was, it took your dinner."

"What!"

"It took the whole Dutch oven."

Marvin was pointing at the ground where the oven had been. The cast iron pot was gone, but something else was there. The same prints at the water hole were in the dust near the campfire.

Johnathan studied the prints carefully.

Then came the ear-shattering scream. It was not like anything that Johnathan had ever heard before, and the sound sent chills up the back of his neck.

Johnathan grabbed Marvin and shoved him in the pickup. He raced around to the other side and started the engine. It was the first time Marvin had ever seen fear in his father's face. Johnathan had become a believer.

"Probably some old hungry hermit," Johnathan mumbled trying to convince himself as they drove away from the site. "We'll come back for the stuff tomorrow. Tonight we go find the sheriff."

The following day the pot was found on the hill overlooking the campsite. It was empty.

CHAPTER 6: SNOW AT SOUTH FORK

THIS NEXT STORY COMES FROM MIKE WHITE. HE TELLS OF AN INCIDENT, WHICH HAPPENED TO HIM WHEN HE WAS A YOUTH ON VACATION WITH HIS FAMILY AT SOUTH FORK ON THE LITTLE COLORADO. HE WAS 12 YEARS OLD AND THE DATE WAS NOVEMBER 22, 1955.

"Do ya think it'll snow?" Mike asked his mother as they loaded the rest of the blankets into the car.

"Maybe," Marsha replied. "It's a bit warm yet, but if the cold front comes in, it could turn to snow."

"I want it to snow." Mike stated with determination. "Will it snow if I pray real hard?"

"Possibly." She smiled. "You can never underestimate the power of prayer."

They finished stuffing the blankets into the back of the station wagon and waited for the rest of the family to come out to the car to prepare for the trip. It would be a nice to spend a vacation with the family in the mountains. Mike's dad had made reservations in a small cabin resort area at South Fork for the weekend. They were going to get away from it all for a few days.

Mike's dad, Jason, was the first out the front door.

"Did you load up the fishin' poles?" Jason asked Mike.

"Yes dad. I even dug some worms out of the garden."

"Good. They've got some good fishing ponds there at the cabins, and there's a creek just down the hill full of trout."

Mike was excited about that. He loved to fish. Something about putting a line in the water was captivating. It really didn't matter if he caught anything, it was more the association with the water that Mike liked. Of course, catching a fish just added to the enjoyment of it all.

The rest of the family came piling out of the door into the station wagon. There were six in all. Mike was the third in order of age with an older brother, Marvin, and an older sister, Dorothy. Then there were the younger sisters, Sue and Marie. And there was the baby boy, Shannon. He was only a few months old, but Mike could hardly wait for him to get old enough to actually do something with him. Actually, he was more waiting for a younger brother to place into slavery much like his older brother had done to him.

The last car door slammed shut and the "leaving for a long trip check list began".

"Did every one go to the bathroom?" Mom questioned. "Yes," came the group reply.

"Honey, did you check the stove and lights?" she asked Jason.

"Affirmative."

"Doors locked?"

"Yes."

"Well, I guess it's time to go," Dorothy stated. She seemed ready to get the trip over with.

Jason started the car, and began to back out of the driveway. The car had barely hit the street when the first interruption came.

" Wait!" Marsha sounded apologetically. "I forgot to check the iron!"

The car came to a stop.

"Just kidding," she laughed nervously. "I haven't used the iron in a week."

Jason smiled and then decided to give everyone a second chance. "Did we pack everything?"

"Yes," came the family reply.

"That's good, cause I don't think we have any more room. We'd have to leave a kid or two behind."

Marvin pointed to Mike. "I volunteer him."

Mike looked mean at his older brother. "No way, Jose. Maybe we ought to leave you."

"Ok," Marsha interrupted. "We're not leaving anyone behind. We have a long way to go so no one picks a fight. Got it?"

"Yes, Mom." came the reply

Jason started the car in motion again.

"Well," Jason joked. "We're off like a flyin' herd of turtles."

"Hey," Marvin jabbed Dorothy in the ribs. "You guys ever hear of the Mogollon Monster?"

"There's no such thing," Mike whispered to the two younger girls.

"There is." Marvin stated flatly. In fact, the last time it was seen was by some Boy Scouts.."

"Really?" Sue squeaked in a small unsure voice.

"Yes. And Boy Scouts wouldn't lie. Would they?"

"Well, no." Mike admitted.

Marvin continued with the story "Last year some boy scouts were out horse riding and they were following the Mogollon Rim just outside of Payson. They decided to camp for the night before going on down the mountain to the main camp."

Marsha and Jason quickly glanced back at the children. It was unusual when they were all cooperating and not fidgeting. Quiet moments like this deserved a closer look, just to make sure there was no mischief a foot.

Marvin continued. "They stopped and pitched camp near a large densely grown area. Soon they had their tents pitched and enough firewood gathered for the night. Because it was going to be a very dark night."

"How'd they know it was going to be a dark night?" Dorothy teased.

"Cause it was overcast and they were in a canyon." Marvin quickly corrected. "And there wasn't going to be a moon."

"Anyway, go on." Marie insisted.

"It was a cold night too. In fact it was started to snow just after they had finished their dinner. They climbed into their tents and went to sleep in their mummy bags. It was so cold that they decided to close the face opening on the bags to keep their noses from freezing. About midnight, they heard footsteps outside their tents."

The two little girls snuggled closer to their big sister. They could already tell this was going to be a spooky story.

"At first the steps were around the tents and outside by the fire place. The scouts thought it might be a bear so they decided to keep quiet in their tents. Soon they could hear kind of a scratching noise on the canvas of the tent. They laid still afraid to move."

"Not me," Mike stated. "I would have been out of there in a heartbeat."

"And you would have gotten eaten," Dorothy stated. "Everyone knows you play dead when a bear is checking you out."

"That's right," Marvin agreed. "Anyway, they were all quiet in their bed rolls when they heard the canvas ripping. They didn't make a sound. One of the Boy Scouts said that he could feel something scratching at his face, but he was covered by the mummy bag. It was almost like it was trying to figure out how to get inside the bed roll."

"I'm scared." Sue whispered to her little sister.

"The scout was dropped back on the ground and then felt like something was picking him up. He opened up a little peek hole in his bedroll and looked out. It was completely black but he could tell he and another scout were being lifted out of the tent. Once outside of the tent, he could see that there was actually three Boy Scouts being carried away. Him and another were being carried, and one was being drug along the ground like a sack of potatoes. He said that the beast really smelled bad."

"Where was it taking them?" Sue anxiously asked.

"He said that it carried them to the bank of a wash and then threw them down the bank and they rolled to the bottom. They laid there for a few moments until they thought is was safe, and they began to open up their bed rolls. By that time they could see a large hairy figure carrying three more scouts to the edge of the bank, and they were tossed over the side like the first three. This continued until all of the scouts were removed from their tents and were deposited at the bottom of the dirt bank. When this was all done, the Mogollon Monster screamed a terrible scream. Then it simply turned and walked away."

"Wow!" Marie exclaimed. "Is that a true story?"

"All of the Scouts say so." Marvin insisted. "They laid at the bottom of the gorge until daylight. Then they braved getting up. All of their horses were gone, and then they decided to take a head count to see if everyone was there."

"Well?" Mike asked.

"All with the exception of one scout that tried to get out of his bedroll and get away. They never saw him again. They never found his body. Some say that the monster kept him cause it thought it hatched him."

"What?" Dorothy insisted.

"Yes, the only thing the others could figure was the monster thought that all of the scouts in the bed rolls were actually a human nest and the bed rolls were eggs. When the scout tried to get out, it thought the egg hatched. It felt like it should keep him since the nest had been destroyed."

Marsha nudged Jason. "What do you think. Why would they tell that story at scout camp?"

"Probably to keep the night raids to a minimum," Jason whispered back to Marsha.

"Oh, I see. To keep them in their beds and out of trouble."

"Yes, that's my guess." Jason admitted. "The story couldn't be true."

The trip was fairly uneventful. There were the usual pit stops. The first restroom break came about halfway between Superior and Globe. One in Globe plus a stop for gas, and one in the Salt River bottom for a restroom break. Another gas stop in Show Low and another restroom break in Springerville.

It was evening when the family finally pulled out of Springerville headed for their final destination at South Fork. It was dark and began to snow.

Mike had been napping but woke up when he heard someone say the word--Snow.

"Yes!" he exclaimed. "It's snowing."

The snow was light, but was getting heavier.

"Do ya think it'll stick?" Mike asked his dad.

"Oh yeah," Jason replied. "There's going to be plenty of it tomorrow."

Mike liked the sound of that. He had never been in the snow and this was going to be a treat. He had seen many pictures of sledding and skiing, but never done it.

"This is going to be a great trip." he thought to himself.

Mike fell asleep with a smile on his face.

The next morning Mike woke up early. Everyone else seemed to be too tired to want to go outside.

"You missed it," Marvin told him. "Last night we had a blow out."

"Really?"

"Yes. You must have slept through the whole thing."

"I did?"

"Yes. Dad and I had a rough time digging us out of the snow bank we skidded into. Then it took us several hours to change the tire in the snow. Dad forgot his gloves, and we could only do so much before it was too cold on our hands."

"Where's Dad now?"

"He took the car back to Springerville to get the tire fixed. He left a few minutes ago."

"When's he gonna be back?"

"He said it would probably be later this afternoon. The roads are packed and the plows haven't been around. He said you could fish the little pond if you want. He'll take you to the creek when he gets back."

With that, Marvin rolled over and went back to sleep.

Mike got dressed and had a bowl of cereal for breakfast. Afterwards, he located his fishing pole and worms. He was ready for the outside.

It was a beautiful, glorious day. The clouds were packed in tight and looked like heavy burnt marshmallows. The snow was still coming down in flurries. But there was no wind. In fact, it was very quiet and warmer than Mike thought it would be.

The small pond was a short walk from the cabin. Mike noticed how it was plainly visible from the front porch window. He knew he would be checked on every now and then by his family, and this was comforting. After all, he was out in the sticks by himself, and it was kind of spooky how quiet it was with the snow falling.

Mike trudged over the snow to the banks of the small pond. There was ice forming on the water, but it wasn't thick enough to support his body weight. After looking around a bit, he located a large rock to which he used to knock a hole in the ice several yards from the bank. A few moments later, Mike had his first worm on the hook and in the water.

The first trout hit a few minutes after the line had hit the water. It was a nice rainbow trout. Mike took it from his hook and stuffed in his coat pocket.

"It's the only logical place to put it," He thought out loud. "Mom won't mind the smell when I bring home enough fish to feed everyone."

Mike baited the hook and tossed it back into the water. He was quite successful. He continued to pull out one fish after another until all of his worms were gone. In total, he had eight fish stuffed into his coat pockets.

"Too bad," he spoke out loud. "Dad couldn't be here. He would have loved catching all of these fish."

Mike packed up his equipment and headed back toward the cabin.

Suddenly he got the strangest feeling run up his back. It was that funny feeling as if he was being watched very closely. He looked at the cabin window, but there was no one standing in sight. Mike peered around at the other cabins in the area. They all seemed lifeless and cold.

It looked as if his family was the only people in the area. He shook off the feeling and started for the cabin once again.

Once inside, Mike peeled off most of his outer clothing and kicked off his boots. He decided to take the fish to the sink in the kitchen and leave them for his dad. He would know how best to clean the fish.

The rest of the family had begun to stir. His mom was in the kitchen cooking up breakfast.

"Good morning," she cheerfully greeted Mike.

"Hi mom."

"I see you caught your limit this morning."

"Yes. They were certainly biting." Mike was still wondering about the feeling he had on his way to the cabin.

"It looks like something is bothering you. What's up?

His mom finished a big stack of pancakes and put them on a plate for Mike. She knew he liked them with lots of butter and syrup.

"Mom, were you watching me a little while ago?"

Marsha slid the pancakes in front of Mike along with the butter and syrup. "I looked at you when I got up about a half an hour ago. Why?"

"Cause I got the creepiest feeling just before I came in the cabin. Like someone was watching me."

Marsha laughed in disbelief. "Maybe it was your brother?"

"No it wasn't me," came an unexpected voice.

It was Marvin. He was busy tying his robe into place as he walked into the room. He still looked a bit tired and strung out.

"I've been asleep until just now. I smelled the pancakes. You got any more?"

Marsha smiled. "You bet."

"Better make that a lot more," came another voice. It was Dorothy, and she was followed by sue and Marie."

"I made plenty of cakes. There will be enough for everyone." Marsha went about doing her best feeding the hungry horde.

"Why would I get the feeling that someone was watching me?" Mike asked.

Marvin put his arm around his little brother. "Maybe it was one of the neighbors?"

"I didn't see any cars at the other cabins," replied Mike.

"Well," Dorothy piped in. "Maybe it's just your imagination or maybe the Mogollon Monster?"

Mike was a little irritated, but the answer was bound to come from one of the family.

"Tell you what," Marvin butted in before Mike could get in a reply to his sister's comment. "After breakfast, I'll go with you and we'll look around. Then you can show me where you caught all the fish. Deal?"

"Sure!" Mike enthusiastically agreed.

Mike was amazed at his big brother's generosity. But, sometimes Marvin was like that. Sometimes his brother would torment Mike until he was ready to cry, but if someone else picked on Mike, Marvin would come to the rescue.

"We want to go too," Sue requested in a statement. She was speaking for her and her little sister, Marie.

"Sure," Marvin agreed. "We'll all go out and play in the snow."

This reply even surprised Marsha.

"Ok, what gives?" she questioned Marvin as he shoveled some pancakes onto his plate in front of him.

"I don't know." Marvin began. "I haven't had much time to play lately."

"That's because you've got your nose in your car engine all of the time," Dorothy teased.

"True," Marvin agreed. "But, I just haven't taken any time out to play in a long time. Besides, I'm eighteen. I'll be on my own soon, and I won't have any time when that happens."

"What's this?" Marsha questioned. "My eldest son is growing up?"

"Not today."

Marvin stuffed some of the cakes into his mouth. "Besides, I don't want my little brother too scared to go outside by himself. He'll want to hang around me all of the time."

"I see." Marsha mused.

After everyone had been fed and dressed for the snow, the group set out for the pond.

Marsha elected to stay in the cabin and watch the group from the window. The party turned up nothing. Mike was right. There were no neighbors. It didn't take long before the children were hard at play. There was a small hill overlooking the area with the pond. It was perfect for sliding down on an old inner tube.

Marvin took a few trips down the slope just to prove he could do it. Afterwards, he was content to stand at the bottom of the hill and fish the pond while he watched the other kids play.

Several hours later the snow began to fall once again. It kept getting heavier as the time passed. Soon all of the kids were exhausted except Mike. He was still busier than ever trudging up the hill with the tube for another trip down the slope.

"Mike," Marvin called. "I'm going to take the other kids back to the cabin."

"Ok," Mike puffed.

"I'm going to warm up and be back out here in a couple of minutes. Ok?"

"Sure."

Mike continued to work his way up the hill for another slide. At the top he turned around and surveyed the area. His brother and sisters had already disappeared into the cabin. He was alone again. It didn't take long for the feeling of being watched to return.

Mike studied the area below to see no movement other than the falling snow. If the feeling wasn't coming from below, then where?

Mike turned slowly studying the woods in a panoramic fashion until he turned a hundred and eighty degrees. He caught a faint smell of something dead. At first he thought it was the fish he had put in his pockets, but the smell was different. There was nothing behind him but densely growing pines. Perhaps this is where the feeling and the smell were being emanated.

Mike studied the thicket closely. But it gave no clues. He did notice the snowflakes where hitting him in the face. That meant he was facing the wind. Whatever smelled was coming from the trees.

Slowly he turned his back to the thickets and prepared to slide the hill one last time. From the bottom he was going to run back to the cabin to stay.

Suddenly a terrible crash sounded behind him. It sounded as if an entire tree was being snapped to the ground. Mike jerked around and faced the sound. In the distance he could see several trees still moving from the falling timber.

The scream came moments later. It sounded much like the screams on TV when a woman was being stabbed or had found a dead body. It sent fear through his body. He dove on to his inner tube and slid down the mountain as fast as he could go.

Once at the bottom, he jumped to his feet and started for the cabin. As he reached the bank of the pond, his foot went sideways from stepping on a rock and down he fell. He was unable to protect his head. He heard the sick thump as his skull hit the frozen ground. It reminded him of the sound when someone drops a bowling ball on the floor.

Mike never did lose consciousness, but he was senseless. He couldn't seem to get his footing. He stood only to fall again and again.

He heard the crack of the ice, and realized he had wandered out over the pond. The next thing he knew he was looking up through the hole in the ice. The water was numbing and cold, but it was too much effort for Mike to try to return to the surface. He was too far gone to care.

Mike lay in the water looking up through the hole in the ice. He thought it was funny how the ice floated on the surface of the water above. But it was soon interrupted. He looked beyond the surface to see a strange, hairy face looking back at him.

The next thing he knew a large hairy arm plunged though he water and grabbed him by the head. He was literally jerked from the water.

Mike lay face down in the snow, still shocked by the events. He felt his body being picked up by the back of the neck and he was being drug through the snow. He thought it was kind of funny how his feet flopped over the steps as he was drug on to the porch.

Moments later, he heard his mother's voice.

"Mike," she frantically shouted. "Do you hear me?"

Mike blinked his eyes and tried to focus.

"Mike!"

He felt his cheeks being patted. Then realized they were being slapped.

"Mom?" He groaned.

"Mike, your going to be Ok." Marvin reassuringly stated.

Mike tried to focus again. This time he succeeded.

"Mom," Mike mumbled. "I...I fell in the pond."

"How'd you do that," Marvin questioned.

"I heard something...Something in the woods. I got scared."

"So you jumped in the pond?" Marvin teased.

"N...No. I fell and hit m...my head."

Marsha pushed Mikes wet hair from his forehead. There as a large bump at the hair line above his right eye.

"My, my..." She stated. "You got a big goose egg on the ole noggin. It's a good thing you were able to climb out of the pond by yourself."

Mike stuttered out the best he could, "I...I...d..didn't... M..Marvin saved me."

Marsha looked at Marvin with surprise then looked back at Mike.

"It wasn't me, bud." Marvin stated. "I found you on the porch on my way back out."

"The...n it's true," Mike mumbled.

He could see the face in his mind, and the hairy arm that retrieved him from the ice water. "Big hairy man grabbed my hair and pulled me out. He drug me to the porch."

Mike shivered from the cold while he forced the rest out. "I..It saved me."

MIKE'S VERSION OF THE MOGOLLON MONSTER

CHAPTER :THE SCARED DOG

THIS INCIDENT TOOK PLACE NEAR GREENS PEAK, SOME THIRTY MILES EAST OF SPRINGERVILLE. LACY JOHNSON TELLS OF THE TIME WHEN THER FAMILY WAS OUT HUNTING AND HAD AN INCIDENT WITH THE MOGOLLON MONSTER. THE EVENT TOOK PLACE ON OCTOER 26, 1962.

It was the last day of deer season and the family had decided to leave the White Mountains early to prevent driving back when they were all tired and worn out. Deer hunting areas were still open to the entire state. The only requirements to hunt were a rifle, hunting license, and a tag. The family agreed to the arrangements and began to break camp.

"We'll hunt our way back," Kent stated to his family. "Since this is our last day and all,"

A short time later they were on the road out of Springerville headed to Mc Nary on highway 260. About twenty miles outside of town they turned north on a trail which would lead them past Greens Peak and Saint Peter's Dome out to Highway 60. This would allow them to hunt some prime deer territory prior to going onto the Apache Indian Reservation where they would have to cease to hunt. There was no game spotted until they had reached the south side of Greens Peak.

"There!" Sarah called.

Sarah, the mom, had always been the eyes of the family. She always seemed to be able to spot the game before anyone else. She pointed out a herd of deer on the hillside nearly a half mile away from the road.

"There's a herd."

Kent stopped the car and pulled out the binoculars to study the heard.

"Yes, there's a couple of bucks in the tree line."

Kent reached inside the car and pulled his rifle from the scabbard. "Sarah, you take the car up the road to the turn off to the top. Wait there, I'll be there in an hour if I don't get anything."

Sarah slid over into the driver's position. "Good luck she called."

Kent started from the car, but was immediately followed by the family dog. It was a cocker spaniel and it loved to go hunting. The only problem was it was too noisy. It was good for flushing birds and rabbits, but definitely too loud for deer.

"Lacy, call your dog back." Kent pleaded. "She can't go this time."

Lacy called her dog, but it ignored her. It was busy trying to find the scent.

"Come here, Lady." Kent called hoping the dog would mind him.

Kent had read that hunting dogs sometimes recognized the human as the leader of the hunting pack and would take its cue from them. He was right. It worked this time. When the dog got close enough, he grabbed her by the collar.

"Keep her in the car. If she gets loose, she'll try to find me."

"Ok, Daddy." Lacy agreed.

Kent turned and was soon out of sight into the trees.

Sarah and Lacy headed up the road in their vehicle.

"Do you think Dad will get one?" Lacy asked.

"If anyone can, your Dad will."

"Do you think Lady can have one of the hooves?"

"Sure."

The dog had pressed its nose through the partial open window. It seemed broken hearted that it had been left behind on the hunt. It would sniff the air and then jump across the back seat to the other window and stick her nose out of the other partially opened window. She seemed more hyperactive than normal.

A few minutes later, they came to the fork in the road. It was the agreed turn off. One road would lead to Highway 60 and the other would take them to the top of Greens Peak where there was a ranger's tower.

"Well," Sarah sighed. "This is where we wait."

"Do you think he's there yet?" Lacy questioned.

"Probably. We should listen for the shots. That might tell us if he got something or not."

"How can you tell that"

"Your Dad carries a special rifle. It sounds like no other. We'll definitely be able to tell when he fires a shot."

"That's how you knew last year to start sharpening the skinning knives?"

"Yes. You see, your uncles and others mostly hunt deer with thirty ought six's. Your dad's rifle is two sixty-four magnum. It sounds more like a cannon."

"But what if he misses?"

"You can tell that too," Sarah replied. "Sometimes when the bullet hits its target, there is a sharp thud after the rifle sounding. If he's close enough, you should be able to hear it."

"But that's more than a mile away." Lacy insisted. "How can we hear that?"

"Sound carries out here in the mountains. Haven't you noticed how you can hear people speaking from a camp across the lake?"

"Oh," Lacy began to understand. "I think it's because there's very little noise out here. There's no noise to get rid of the other."

"You might be right." Sarah agreed.

The two girls were interrupted by the dog. It was not content to stay in the back seat. It was now trying to get out of the windows in the front of the window.

"What's with this dog?" Sarah grumbled as she shoved the dog back from her. Lacy tried to calm the dog down by holding it back from the window.

"I think she wants out bad."

"I've never seen her like that," Sarah agreed. "Put her leash on and tie it around your wrist. We'll see if she needs to use the bushes."

Lacy did what her mother had requested and opened the door. The dog shot out on the run until the leash jerked tight. The dog coughed from the sudden strangling jerk, but it didn't give up. She pulled even harder almost dragging Lacy from the car.

"What gives Lady?" She asked her dog. "What is it?"

The dog wined and sniffed the air.

"Mom do you smell something?" Lacy asked Sarah.

"Kind of," She replied. "Maybe it's a skunk."

"I don't think so." Lacy replied.

The dog began to growl lowly.

"Whatever it is," Sarah spoke softly. "It's close by. Maybe we ought to get back into the car."

"Are we in danger?" Lacy looked at her mother's face.

"No. We just don't want to take any chances. The animal out here might have rabies."

The two started back into the car, but Lady would not budge.

"Rabies?" Lacy grunted pulling on the dog's leash.

Sarah came around the car to help put the dog back inside.

"Yes. Sometimes when an animal isn't afraid of humans or campfires, they have rabies."

"Oh," Lacy grunted when she pulled harder on the leash.

Suddenly the leash snapped. The sudden release of pressure threw Lacy flying into the car seat. The dog took off running into the thicket.

"Lady!" Sarah screamed. "You get back here!"

The dog was gone.

Lacy climbed back out of the car and looked at her mother.

"Sorry Mom. The leash broke."

Lacy was almost in tears. Sarah patted her daughter on the head and nudged her toward the car.

"That's Ok," she spoke softly. Just get back in the car. We'll let Kent find her when he gets back."

The two climbed into the car, locked the doors and waited.

Moments later the two could hear their dog growling. It sounded as if the dog was only a hundred yards or so from the car, but in a terrible fix.

"Mom," Lacy cried. "Lady is in trouble."

"Yes, it sounds like she found the animal."

Lacy was very concerned for the safety of her dog. "I hope it's not a porcupine."

"Me too."

The conversation was interrupted by a loud scream. Was it a mountain lion or something else. Whatever it was, it sent chills down their spines.

"What was that?" Lacy slid closer to her mother.

"I don't know." Sarah was searching the tree line through the windows. There was a faint sound in the distance, but it was coming closer to the car. It sounded like something heavy was being drug through the forest. Then some movement could be seen in the tree line.

"I think its Lady," Sarah stated quietly. "I think she's dragging something."

The two watched more closely at the coming movement.

"Lacy, stay here."

Sarah opened the door and ran quickly towards the dog. As he got closer to the animal, she could see the object being drug was a large log. It seemed impossible that a log so large could be drug so quickly by such a small dog. Sarah was completely amazed. The dog's leash had become entangled in some branches. Quickly, she unhooked the leash and grabbed the dog in her arms and headed back towards the car.

Once inside, she started checking the dog over for damage. There was no outward physical damage, but the dog remained absolutely still. It seemed to be unconscious.

"I think the dog passed out." Sarah nervously commented.

"Will she be all right?" Lacy begged.

"I think so. There doesn't seem to be any bites or broken bones. But, we'll need to get her to the vet."

A resounding crash came from the hood of the car. The sound caused the two girls to scream in fright. Before they could realize what had happened another rock hit the top of the car and rolled off.

Sarah reached for the keys to the car, and it seemed like an eternity before the engine started.

Another rock struck the car's hood leaving a deep dent in the metal.

Sarah slammed the car into gear and hit the gas. The vehicle lurched forward spinning the tires. She swung the steering wheel hard to the left causing the car to swing around onto the road to which they had came.

They had barely traveled a hundred yards down the road when Sarah spotted the figure standing in the middle of the road. She hit the brakes bringing the car to a sliding halt. The figure was Kent.

"Get in the car!" She shouted.

"What's going on?" Kent demanded.

"Get in the car Kent!" Sarah pleaded.

As he opened the car door, Kent spotted the dents in the hood and the top of the car.

"Who did this?" He demanded.

Sarah grabbed Kent by the shirt and pulled him into the car. "Get us out of here!

Now!

A rock slammed into the car at the spot Kent had been standing. He looked at the rock in disbelief. A rock that large would have killed him. He quickly glanced around. They were a hundred yards from any trees. The rock had to have been thrown from a greater distance than that.

Kent slammed the car into gear and squealed rubber from the site. He would never hunt Greens Peak again.

CHAPTER 8: THE FLAGSTAFF INCIDENT

SO FAR IN OUR STORIES THERE HAVE ONLY BEEN A FEW LIGHT CASUALTIES. SOME CATTLE, A COUPLE OF DOGS, AND A GENTLEMAN WHO WAS SCARED SO BADLY HE COULD NO LONGER SPEAK. IN THIS INCIDENT, IT IS NOT SO. THE INCIDENT CAN BE TRACED THROUGH NEWSPAPER ARTICLES AS AN ATTACK BY A CRAZED HERMIT OR LARGE BEAR. IT HAPPENED JUST SOUTH OF FLAGSTAFF ON HIGHWAY 17 NEAR STONEMAN LAKE. IT HAPPENED IN AUGUST, 1963 TO A FAMILY ON VACATION HEADED FOR THE GRAND CANYON.

"Come on everybody," Janet called to her children. "We're not going to get out of here before it gets hot."

John, her husband, was the first out of the house with an arm full of sleeping bags. He brushed by her and paused long enough to give her a quick kiss.

"Honey," he smiled. "This is going to be a vacation to remember."

Janet smiled back. "Do you think we can make it out of the valley before it gets hot?"

"Not a chance." John replied. "It will take us all day to get on the other side of Phoenix."

Janet grimmest the thought. She wasn't too happy about traveling long distances in a car without air conditioning. The centrifugal window cooler wasn't working either, the evaporative pads were shot. But, at least the car was in good condition. The car was a 1963 Chevy station wagon. They owned the car since it was new, but she wanted to trade it in on a new car with air conditioning for the Arizona heat.

John rushed by her to grab another load of camping gear.

"We'll buy a new car next year." John stated as he passed by.

It was almost as if he had read her mind. But that wasn't too hard to do at the moment. Janet wasn't happy about spending all of the saved new car money on this last minute vacation trip to the Grand Canyon. She agreed to go on the trip because John had been working too hard and needed a break from the job. He also needed to spend more time with her and the kids.

Finally, the kids began to appear carrying all of the stuff they wanted to take with them on the trip. Steve, the oldest was the first out the door. He was fourteen years old, and a typical teenager. His arms were full of comic books, all of which he intended to read alone. He was a good kid. He just hated sharing his comic books with his brother and sister.

The youngest child was the next out the door. This was Mikie at the grand age of eight years old. He was dragging his favorite toys along with him. He had his two cap six-shooters and his "Rifleman" Gun. It was a plastic replica of the rifle used in the TV series the Rifleman.

"So what are you going to do with those?" Steve teased.

Mikie stuffed the toy guns into the back seat and turned to Steve. "I'm going to shoot any bears or mountain lions we might see."

"Oh." Steve mumbled with satisfaction. "Protection Uh?"

"Yup."

The boys climbed into the back seat and awaited the rest of the family.

Carolyn was the last child. She was the middle of the two boys in age and often caught the brunt of their warning back and forth. However, she could stand her own. She was quiet capable of wrestling either boy to the ground when she felt the need to do so. She for her toys, she had brought a book of cut out paper dolls and several books from the library. She climbed into the back seat next to her brothers and began to read.

"Ok, this is the last of it." John announced triumphantly as he came out the front door.

He had his hands full of pillows and a small ice chest.

"I made a bed in the back so the kids could go back there and sleep if they wanted to." John told Janet as she sat down in the car. "And I brought this for you."

John placed the small cooler at Janet's feet. She opened it up and found several Barq's Root-beer cooling in ice. They were her favorite.

"That's nice," she smiled for the first time. "This is going to be a good trip."

John was bound and determined to make this trip pleasant for Janet. He knew she was disappointed about the car situation, but he had just scored his first smile. A few minutes later, they were on the road.

They traveled up Main Street through Mesa until Mill Avenue. There they turned north over the Tempe Bridge and over to Scottsdale Road.

"I hear they are going to put in a freeway soon." John commented.

"Oh?" Janet replied. "Where?"

"Its supposed to run north and south through Phoenix over by nineteenth avenue then turn east by Broadway. It will turn south before it gets to Mesa, and head for Tucson."

"Wow. That will knock off several hours off when we go to Prescott."

"Sure will. Too bad it isn't there now."

Janet smiled again. Jason was on a role.

Finally the family reached the north side of Phoenix. It had been a hot day. In fact it had been an extremely hot and dry summer. They all agreed that the pine trees were the place to be.

It was a long climb out of Black Canyon. The car had overheated twice, but Jason had been able to cool it down with the water in the desert bag hung on the front of the hood. This had cost them time, and they were not going to make Flagstaff by dark.

"Well," Jason suggested. "How about we stop up here a ways and camp out for the evening."

"That sound good," Janet agreed. "I don't want the kids to get too cranky from traveling all day and night."

"I know a place where we can pull off the road a ways and camp."

"How far?"

"Just a few minutes more."

Shortly, the car pulled off on to a dirt road. A half mile later they found a good campsite.

"Ok kids," Jason called. "You all know the drill. Get some fire wood."

The kids set about their task collecting all of the sticks in the area. Jason pulled out the tent and had it pitched in only a half an hour. Jason was grumbling something under his breath.

"What did you say?" Janet asked as he brought over the bedrolls.

"Oh, Nothing." Jason grumbled. "Someone needs to design better tents. That's all. Something more light weight than canvas. And something that doesn't have a pole down in the middle of the floor."

Janet tossed the bedrolls into the tent. "You know what they say about tents don't you?"

"No. What?"

"When the box says it's an eight man tent, that doesn't mean that it will sleep eight men."

"Oh...Really." Jason mumbled socialistically. "What does it mean?"

"It means that's how many men it takes to put it up!" Janet laughed.

Jason smirked, then broke out laughing. "That's got to be true."

Janet looked inside the tent. "Honey."

"What."

"The tent isn't big enough for all of us."

"True." Jason snickered.

"Well, where are we going to sleep?"

"Over here." Jason pointed at the station wagon. Jason walked over to the wagon and climbed in the back seat to release the back of the seat. It folded down almost flat.

"I didn't know the car could do that." Janet was surprised.

"That's because I fixed it. All it took was some hack sawing, bolts, and a couple of gate latches."

"That's neat!" She exclaimed.

"Someday, all cars will have fold down seats." Jason boasted. "Here"

Jason tossed a box to Janet. It was a blow up mattress.

"You want me to blow this thing up? Janet questioned.

"No. Just take it out of the box and spread it out in the back of the car like you want it. I'll do the rest."

Janet went about the task. When she was done, she found Jason doing something under the hood of the car. He finished up and strung a tube out and hooked it to the mattress.

He then walked around to the driver's side and climbed in. He started the car. Instantly, the mattress began to swell. He revved the engine a couple of times and shut it off. The mattress was full.

"That's amazing!" Janet stated in astonishment. "How'd you do that?"

"I call it my emergency in flater . I took an old spark plug and knocked the porcelain out of it. Then I welded a piece of tubing to it and attached the hose. The compressed air comes from the engine. I just have to put the real spark plug back in when we are done."

"That's really neat." Janet stated.

"There's only one drawback," Jason admitted. "You can't smoke in bed."

"I don't smoke," Janet looked puzzled. "You know that."

"It's a joke," Jason apologized. "The air blown into the tube has gas mixed into it from the carburetor. Too many sparks could make it blow."

"We'll just have to remember that won't we?" Janet playfully teased.

"Yes. Yes, we will." Jason smiled.

"We finished the wood," Steve interrupted. "Can I start the fire now?"

"Sure." Jason agreed.

"I'll get the hotdogs and marshmallows," Janet announced.

Jason looked at the other two kids. "I guess, we'll go cut some sticks. Right?"

"Right." Mikie and Carolyn reported like soldiers.

Several hours later, the family had finished their marshmallows and stories and decided to retire for the evening. The children were put in their bedrolls and given a flashlight for any of the middle of the night trips to the bushes. Jason and Janet were soon sound asleep in their new car bed.

Steve woke up to a sound outside of the tent. It sounded like heavy breathing and footsteps.

"Carolyn," Mike quietly whispered as he shook her arm. "Wake up."

"Wha...t do you want?" Carolyn grumbled still half asleep.

"There's something outside."

Carolyn's eyes popped open. "What is it?"

"I don't know. Whatever it is it's big."

"Could it be a bear?"

"I don't think so. Mikie, wake up."

Mikie was already awake and was hiding in the bottom of his bedroll.

"What?" he called quietly.

"There's something out there." Steve replied under his breath.

"I know. I saw it. Its shadow was on the tent."

"Well, what is it?" Carolyn insisted.

"It's a big man. Stinks too."

The two were now almost as afraid as Mikie.

"Should we yell for help?" Carolyn questioned.

"No." Steve replied. "He's probably scrounging for food. He'll leave when he finds none."

"What if he doesn't?" Carolyn insisted.

"If we scare him, he might hurt us. Just keep quiet."

The children huddled together in fear listening to the sounds outside their tent.

"Why doesn't Mom and Dad wake up?" Mikie finally spoke.

"Cause they're in the car. They don't hear it."

Suddenly something hit the tent. The children saw the shadow moving along the side of the tent. It appeared to be a large man with long hair and a long beard. It looked as if it was looking for a way into the tent. No one uttered a sound. Steve fumbled around on the floor for something to use as a weapon. He found Mikie's play rifle. He had an idea. Quickly he moved to the door of the tent.

"Who's out there," Steve spoke in the lowest adult voice he could generate. "I've got a gun and will shoot."

The figure stopped in its tracks near the edge of the tent, but uttered no sound.

Steve stuck the barrel of the toy gun out of the door.

"I've got a gun and I will shoot!"

The figure turned and ran away from the tent. All of the sounds indicated it was back in the forest.

The tree kids rushed the car and began to beat on the windows.

"What's going on out there?" Jason spoke.

"Mom, Dad." Steve called through the window. "There's a burglar out here."

Jason drew his pistol and flashlight from under the seat and began to search the area. As he climbed out of the car, the kids climbed in. The family sat in the car watching Jason and listening to every noise in the forest. Jason finally came back to the car and climbed into the front seat.

"It will be light soon," He commented. "I'll stand watch till morning. We'll get our stuff in the morning and get out of here."

"Jason what is it?" Janet could sense something was wrong.

"There's some prints over by the tent." Mike stated grimly.

"So what is it?"

"They look human."

"Well cows don't wear shoes." Janet insisted.

"The foot prints are bare footed."

"That is strange. Maybe it was a bum looking to steal some shoes."

"I don't think so." Jason shook his head in the negative. "They don't make shoes that big."

The family sat in the car until it was light enough to see. They quickly broke camp and headed into Flagstaff. When they arrived, they found a gas station to fill up.

"You're our first customer for the day," The station attendant cheerfully greeted them.

"Can I fill 'er up?"

"Regular please." Jason requested.

"You folks drive all night?"

"No." Jason replied.

"You didn't stop down by Stoneman Lake turn off did ya?"

"Yes. Why?"

"Did ya see anything funny?"

"Well, we had someone visit our camp in the middle of the night."

"And you stayed there?"

"Yes."

"Well you folks are either very brave or stupid. Maybe both."

Jason looked a bit perturbed at the comment. "What do you mean?

"Well, there was another attack last night."

"Another?"

"Yes. A truck driver was pummeled to death, and his wife was drug off into the woods."

Jason was shocked. "Did they find the woman?"

"Yes, she managed to escape. She was badly brutalized and had several large claw marks down her back and legs."

"Have they caught the guy?"

"Funny you'd say that. She said it wasn't a man. She insisted it was a large ape like creature with long hair."

"You're kidding."

"Nope. But the police are looking for a large man they think might be a crazy hermit. They said the wounds on the truck driver had to have been made by a large club with some type of spikes driven in it. But they still can't explain how he was so badly torn apart."

The station attendant finished pumping the gas. "That'll be two dollars."

Jason paid the man and climbed back into the car.

69

"You believe that?" Janet asked.

"I don't know. But we'll go home by way of Payson just to make sure."

CHAPTER 9: THE STALKING AT BEAR FLATS

MAN HAS ALWAYS BEEN THE SUPREME PREDATOR IN THE FORESTS, AND MANY HUNTERS SHOW NO FEAR WHEN THEY STARE DOWN THE BARREL OF A GUN AT A DEER, ELK, OR EVEN AN ELEPHANT. BUT WHAT HAPPENS WHEN THE HUNTER FINDS OUT HE IS THE HUNTED? WHEN THE HUNTER HAS BEEN STALKED, HUNTED, OR TRAPPED BY A SUPERIOR PREDATOR AND THE JAWS OF DEATH ARE ABOUT TO SURROUND HIM? IT WAS ONCE SAID, "SOME DAYS YOU GET THE LION, AND SOME DAYS THE LION GETS YOU. THESE NEXT FEW STORIES LET US KNOW THAT HUMANS ARE A PART OF THE FOOD CHAIN AND SOMETIMES WE ARE NOT AT THE TOP. THIS STORY COMES FROM LENNY WILSON. IT WAS HIS FIRST DEER HUNT IN AN AREA ALONG THE MOGOLLON RIM KNOWN AS BEAR FLATS JUST EAST OF PAYSON. THE INCIDENT TOOK PLACE IN OCTOBER, 1965

"It is a beautiful day," Jack groaned as he stretched his body in front of the tent. He looked to the east and could see the horizon growing lighter with each passing moment.

"But it isn't even daylight," came the reply from the tent.

"Come on boy," Jack insisted. "We need to get breakfast and get out there before daylight."

Lenny climbed out of his bedroll and slipped on his shoes. He was glad he had decided to sleep in his clothes because there was a chill in the air. He was amazed at the sight of his breath in the cold air.

"Dad?" Lenny called from the tent door. "Where's Jeremy?"

Jack looked around the camp and didn't see his eldest son. "He probably went for more wood or something. I'm sure he'll be back when he smells breakfast."

Lenny walked over to where his father was trying to light a fire. The pine needle were starting to catch, but there was still a lot of smoke. Soon the needles were burning and the twigs were catching. They would have a good fire in a few moments.

"Feels good," Lenny told his dad while he waved his hands over the flames.

Jack stood and walked over to the truck retrieving a large frying pan, the spatula, and ice chest. He carried it back over to the spot next to the fire.

"Here," Jack handed Lenny the frying pan. "Make yourself more useful."

Lenny propped the pan up on the rocks to where the flames of the fire would heat the metal. His Dad handed him a cloth and the cooking oil. This was a procedure to which Lenny was quite familiar. Whenever they had a dry camp, they would not be able to wash the pan very well. To clean it they would heat the pan until the previous cooking oils would start to sizzle. Then they would wipe it clean with a cloth. The procedure would be repeated by putting some fresh oil in the pan and get it good and hot and wipe the pan clean again. This would serve to clean and sterilize the pan for the next cooking.

Lenny finished the chore and put the pan on the rocks to reheat it for cooking. His dad popped some bacon on to the hot metal and there was an instant hiss. Lenny liked the sound. In fact, he loved the whole idea of camping. He often felt he was born a hundred years too late because he loved the outdoors so much.

Jack looked up at the ever-brightening horizon. It was a beautiful sight. The sky was turning a orange yellow, and there was a hint of change in the air.

"It will be bright enough to hunt in about a half an hour." He spoke softly. "We'll need to hurry breakfast."

Lenny heard something in the bushes. He perked up trying to see what it was. A few moments later, his brother appeared with an arm full of wood.

"I smelled the bacon," he commented. Jeremy dumped the wood onto the dwindling woodpile.

"We havin' eggs or pancakes?"

"Eggs," Jack replied. He turned to Lenny and smiled, "Told ya. He'd be back when he smelled breakfast."

"Dad," Jeremy spoke in a decisive tone. "I've been thinking."

"Ok son," Jack stated with a straight face.

Jeremy continued, " I think I'll work my way down the edge of the canyon to the north for about five miles. Then circle back following the creek. I spotted some large tracks headed that way. I'll bet it's that big ole buck everyone has been looking for."

"Sounds good," Jack stated. "I think I'll hunt to the south towards Hell's Gate, and Lenny can go east."

Lenny almost choked on a piece of bacon he had illegally snatched from the frying pan.

"Really?" He gasped. "I get to hunt by myself?"

This time it was Jeremy and Jack laughing.

"Yes." His dad reassured him. "You're old enough to hunt on your own."

Lenny jumped up and started to look around the camp for his gear to take on the hunt.

"Hold on son," His father called. "Let's finish breakfast before we take off."

Lenny realized he was not behaving like an old experienced hunter. He regained his composure and walked back to the fire for breakfast. But inside, he wanted to be on his way.

A short time after breakfast, the three split up going to their designated areas. Lenny was immediately on the hunt. He walked as quietly as possible stopping every now and then to watch the area for any movement. He did this for several hours until he was convinced there was no game in the area. He had only made it a mile or so from camp.

"Perhaps," he thought to himself. "I'm going too slow. Maybe if I go a little faster, I can cover more area. Besides, deer have to make noise when they walk. Maybe they'll think it's just another deer coming their way."

Lenny picked up the pace.

Three hours later, Lenny noticed the brush was getting thicker. The area seemed to be losing the trees and more underbrush was everywhere. Soon, he found himself fighting the brush just to get to a rock point where he could see what was ahead. When he climbed out on to the rock outcropping, he was very discouraged. As far as he could see, there was nothing but a sea of thick scrub oak and Manzanita brush. There would be no way to effectively hunt the area. He decided to backtrack several miles to the trees and to stay out of the brush.

A short distance from the rocks, Lenny thought he heard something in the brush. He stopped to listen. There were no sounds.

He began to pick his way through the brush again but came to a halt when he heard a twig snap behind him. Quickly, he swung his rifle from his shoulder and studied the area. There was no further sound or movement.

"Must be a skunk," He told himself. He had heard stories of people often followed by skunks. The skunks were a curious animal, and would sometimes follow hikers in hopes to pick up some scraps of food.

Lenny began to walk again following his back trail towards the trees. This time listening for any tale-tell signs of him being followed. He had to swallow hard when he heard the first footsteps.

At first, he noticed the steps came at the same rate as his. Whatever it was, it was trying to match its steps with his. When he stopped, so did the noise.

Lenny began to sweat. He wondered if it was a mountain lion or a bear. If so, how far away was it? Lenny could only see a couple yards in any direction due to the thick brush. He needed more visibility.

Slowly, Lenny moved through the brush watching his back as much as he could. There was no sound following him. Finally he worked his way to the edge of a slope, and below he could see a large clearing.

"If I can get to that clearing, I'll be safe." He told himself. "It will have to show itself to get to me there."

Lenny started down the game trail that seemed to lead to the clearing. The sound of a cracking twig rang in Lenny's ears. Whatever it was it had moved closer without a sound. It was stalking him!

Lenny broke out in a run for the clearing. He was running for his life. Behind him the bushes came alive. It too was on the run, hot on Lenny's trail.

While running, Lenny managed to open the bolt of his rifle and to shove a shell into the chamber. The whole time he was hoping to make the clearing before having to use it. The footsteps were coming closer. They were heavy and much further apart than Lenny's stride. It had to be a huge animal of tremendous weight.

Suddenly, Lenny fell to the ground. He had tripped over a large white object in the trail. Fortunately, Lenny didn't lose his rifle in the fall, and he had managed to keep from hitting his head on some rather sharp looking rocks. Six inches taller and he would have been seriously hurt. Lenny rolled over and prepared to fire at the on-coming rush. But it never came. It had stopped short of being visible in the brush.

His heart was beating wildly, and sweat was rolling down his face into his eyes. The sweat stung and burned. He searched the back trail for any movement. Then his eyes focused on the object that had tripped him. It was a bone. It appeared to be a leg bone of some sort, and it was about eighteen inches long. It appeared to have been propped up in the trail with some rocks to create a formidable tripping hazard. It had done its job well, but Lenny had broken it in his fall.

Then a terrible thought hit him. He wasn't being stalked by an animal. This was a trap. Someone meant to hurt or even kill him. But who? And why? He was being herded down the trail to what he thought was a place of safety. The whole idea was to get him running down the trail to trip in the trap. The bait was the safety of the clearing.

Who or whatever was doing this, knew how the human mind worked. But was the trap meant to kill or disarm him.

Lenny searched the bushes with his eyes for clues. A short ways away he could see what appeared to be a pile of bones. The white pile glistened in sunlight that had filtered though the leaves.

He remembered how years before he had been visiting his cousin's ranch in northern Arizona. The hills were low and rolling between the old extinct volcanoes between Springerville and Saint Johns. In these rolling hills, they found some rock walls. These walls were about six feet tall and about the same in thickness. They seemed to be coming from nowhere and going nowhere. They had followed the wall for several miles when they spotted another wall, which seemed to be merging with the first. Nearly a half mile further east the two walls merged, but did not touch. They created a narrow corridor then ended into the open plains again. At the opening, they found several pits. These pits were square in nature and were lined with rocks. Each of the pits looked as if someone had taken a pyramid and made a hole in the ground with it. When their uncle had caught up with them, he explained that the walls were part of an elaborate ancient trap.

Prehistoric men would get the animals running using the walls to funnel the animals into he final stages of the trap. The only thing that kept the animals from leaving the funnel was they could always see a way out in front of them. The pits were invisible. When an animal fell into the pit it would not be able to touch bottom. One wall would catch the front of the body and the back wall would catch the rear snapping the animal's spine doing the hunter's dirty work.

Lenny slowly raised to his feet and backed his way down the trail to the clearing. The last remaining yards he turned and ran into the center.

A few minutes later, could hear the bushes moving. He could actually see the bushes parting as the beast passed through them. However, he could not see the beast itself.

"Whoever you are, I will shot!" He called out nervously,

There was no reply. The animal was now circling the clearing looking for an advantage.

"If you are human, you better speak up!" Lenny demanded.

Still no answer came.

Lenny lowered his 7.7 mm Japanese rifle towards the noise in the bushes.

"Last chance! Answer now or I'll blow you away!"

Lenny took dead aim at a moving brush and squeezed the trigger. The rifle leapt in his hands with a defining roar.

An agonizing scream sounded from the bushes. The scream sent chills running up Lenny's back. It was not like anything he had ever heard before.

Lenny shucked the spent casing and loaded another. He prepared to fire again. As he pulled up the rifle, he could see the bushes moving away from him. The beast was making a beeline away from the clearing. The bushes seemed to offer no resistance as it moved through them. No animal he had ever seen could move through the bushes like this beast.

Lenny sat down and composed himself. How was he going to get out of there before nightfall? Then he remembered his father's training, "Fire three shots, wait a few moments and fire three more. That's the hunter's distress signal. If we hear you, we'll fire four in return and we'll come a runnin'."

Lenny counted his shells. He had nineteen more. He fired three shots and then three more. He waited and listened, but no return. He felt as if he was completely alone with the beast out there waiting for another chance.

An hour passed, and Lenny tried the signal again, and still no reply.

Lenny knew he had to get closer to camp or his family would never hear him. He counted his bullets once again. He had used thirteen. Having enough to try one more time, he decided wait until he could get closer and try again an hour before dusk. Noise seemed to travel better at dusk, and his family would be listening for any distress signals that late in the day. If there were no answer, he would be on his own and have to try to make it back to camp. He would save the last two bullets for the beast if it returned.

Lenny started up the trail from which he had come. He moved as quietly and swiftly as possible. At the top of the rise, he looked back on to the clearing. He was watching for any movement below. Moments later, he was moving up the trail which had lead him into the dense under brush. He was making good time and still no signs of the monster.

Monster?

Lenny though back to when he was at Scout Camp. He thought of the stories of the Mogollon Monster and how they sat around the campfire scaring each other until they could no longer sleep at night.

He laughed nervously as he thought, "We were such kids then."

Finally, he reached the top of a ridge overlooking the campsite below. He had less than a mile to go, but he was exhausted. Now was the time to fire the three shots.

He fired the first volley and waited for a few moments. He fired the second.

Suddenly there was movement in the buses behind him. His heart seemed to fail and everything seemed to go into slow motion. In reality, it only took him a moment to turn, but it seemed like an eternity to him. With his rifle ready to shoot to kill, Lenny drew down on a large four-point buck.

A sudden rush of relief swept over him as he lowered his gun and watched it run down the trail. He had lost the desire to hunt.

In the distance, four shots rang out. His family was on the way.

CHAPTER 10: THE MISSING BUCK

THIS STORY TAKES US BACK TO THE ESCUDILLA MOUNTAIN AREA ONCE AGAIN SOME 50 YEARS AFTER THE MOUNTAIN WAS BURNED. DAVID CASBURN TELLS OF AN INCIDENT BOTH HE AND HIS BROTHER-IN-LAW WILL NEVER FORGET. IT HAPPENED NOVEMBER, 1975.

The morning was cool and the air felt crisp. It had been snowing most of the morning, but it looked as if it was going to clear up.

David Casburn and John Smith left their pickup and started working their way up a long draw, which would eventually lead them to the top of Escudilla Mountain from the south side. They were hoping to get at least one shot at a deer before hunting season was over. The trip would take most of the day.

David decided to drop down into the draw and work the thicker brush while John stayed high to hopefully get a shot at anything David flushed out. It seemed like a plan that would surely work. However, they were soon to find out they were mistaken.

They worked their way up the draw not seeing any deer. This was unusual because they had spotted deer every time they had hunted the draw. Eventually the going in the canyon got to be too tough for David so he started his climb out to find his brother-in-law.

As he approached the top, he could hear voices. John was talking to a group of hunters coming off the side of the mountain. David stumbled over a rock, which made a noise. The other hunters grabbed their weapons as if they were under attack.

"Ho there!" David raised his hands in the air to signal he meant them no harm.

"Gentlemen," John spoke. "This is my brother-in-law, David Casburn."

The other hunters put down their rifles and relaxed a bit. They each offered their handshake in an apologetic manner.

"Sorry," One of the spoke. "I guess we are a bit skittish."

"Oh?" David attempted to get more information from he hunter.

"You won't see anything in the canyon," the gentleman continued. "We already swept it this morning."

"Well, that explains it." David commented to John. "We might as well head back to the truck and find another place."

"Well," the eldest hunter interrupted. "You're welcome to the buck we shot a few hours ago."

David and John looked at the older hunter in surprise. "You shot a buck?" David questioned. "Where is it?"

The older hunter pointed out a prominent outcropping of rock on the hillside. "It will be in the thicket below that point of rock."

"Don't you fellas want it?" John asked.

"No." The older man insisted. "It would be too much effort for use to retrieve."

"Did you tag the animal?" David questioned.

"No. We never actually found the deer."

Both David and John thought this was peculiar. Why did these hunters shoot a buck and not retrieve it.

"So you fellas headed out of here?" John asked.

"That's right." The older gentleman stated. "We're headed back to the valley."

"Then we'll go try to find the deer." David replied. "You sure you don't want us to help you retrieve it."

"No." Came a flat answer from the youngest of the group. "We'll be leaving now."

John and David watched the group of hunters hurry down the hill side.

"You know," John started. "At the rate they're going, they'll be back to the road in an
hour."

"I know." David replied. "Something sure spooked them."

"Do we go after the deer?

"Yes, we do. We're not going to leave perfectly good meat to spoil on the mountain side because some city folks are too scared to go in a thicket an get it."

"We can be up there in about an hour." John replied. "That leaves us about three hours to look for the deer and get back to camp. Let's do it."

The two headed up the mountain.

"You ever heard of the Mogollon Monster?" John asked.

"Yes." David admitted. "I've heard a few stories of it, but it's just a scary campfire story."

"Well, it's a story that has been around for many years."

"What do you mean?"
John continued. "The Boy Scout stories took place over near Payson. But, that isn't the only place it has been seen."

"Oh?" Mike smiled. He knew his brother-in-law was trying to scare him. "Where else?"

"Here."

"No kidding," Mike stated socialistically.

"I'm not kidding," John continued. "Have you noticed the old burn rubble here on the mountain?"

"Yes."

"The locals burned it to kill off the monster."

"Monster!" David laughed. "How long ago was that?"

"It was in the early nineteen hundreds."

"Come on. That would put the monster over fifty years older than it was back then. If it's still around it would need a walker."

"You don't think it could live that long? Well, what if there was more than one?" John insisted.

"Not a chance. They would be a trophy on someone's wall."

"Possibly," John admitted. "But, you have seen very smart deer, right?"

"True."

"They know how to evade most hunters. That's how the big ones get big."

"So what's your point."

"What if they are intelligent primates? Maybe they know man means death and they avoid man as much as possible?"

"Ok, even if it was so. Someone would have found some remains from one that died."

"Maybe not." John insisted. "Maybe they eat their dead."

"What?" David stopped and looked at John. "Apes and gorillas are not carnivores."

"Who says? Anyway, it is a natural thing for animals to eat their dead. Cats do. I mean at birth. If a kitten is born dead the mother will eat it."

"Gross, John. That's gross."

"There are others that will eat their young and dying."

"I suppose. But there would be bones."

"Not if they were scattered over a large area. They would look like any other dead animal."

"No animal does that." David stated.

John disagreed. "Oh, not so. Elephants have been known to mourn over their dead and then carry off the bones for many miles."

"Really. So what brought about this conversation anyway." Mike started up the hill again.

"Something scared those other hunters. So much that they gave up the hunt on a dying animal."

"They probably seen a bear or something. We can handle that."

"True." John admitted. "But, let us be careful."

"Sure." David agreed.

The two walked along not saying much. Once at the thicket, it didn't take them long to find the blood trail of the wounded deer. The snow made it easy to track. And, it was quite obvious, that the three hunters had made it this far. Their tracks were paralleling that of the deer's trail.

About a hundred feet into the woods, they found where the deer laid down in the snow and thrashed around. It looked as if it was preparing to die.

"It won't be long," David whispered to John. "It can't have gone too far."

"This is where the hunters turned around." John replied. "They went downhill for some reason."

They picked up the deer's trail again. The deer had staggered and stumbled for another twenty-five yards and laid down thrashing around again. The blood loss was tremendous.

"Something is driving the deer," John whispered to David. "Otherwise, it would have died by now."

"It looks that way, but I don't see no hunter's tracks."

"I don't either." John replied. "Something's weird."

They found the trail again and followed it deeper into the thicket. Again they found where the deer laid down to die.

"This should be it." David stated.

They looked around. The snow was trampled as if there was a small battle fought on location. Blood was everywhere, but there was no buck.

"Where could it be?" John questioned. "It didn't grow wings and fly off."

"Let's try to pick up the trail again," David suggested. "I'll go out twenty five yards and circle. There's got to be some sort of tracks."

John stayed to study the area a little longer. David circled the entire site and came back to the center following his own tracks.

"There's no tracks leaving this site." David told John.

"That's impossible. The buck is here somewhere."

The two searched the area once again. This time they both walked the circle looking for tracks.

"We're running out of time." John stated. "We're going to have to give it up."

John shook his head in disbelief. "Out smarted by a mortally wounded deer. How can that be."

The two made their way back to the last thrashing site. John stopped to rest a moment and leaned against a tree.

"David, is there anything in these woods that could lift a deer into the trees?" John questioned.

"I don't think so. A mountain lion might be able to carry a small deer, but not a big buck. A bear wouldn't worry about it. It would bury the carcass and come back to eat on it until it was gone."

"Well," John spoke kind of funny. "Explain this."

John pulled his hand from the tree. It was covered with blood.

"Unless this deer learned to climb a tree, something took it up there."

Both looked slowly up the tree following the blood trail. It was impossible through all the thick branches.

"I'm for getting out of here." David whispered to John.

"Me too."

CHAPTER 11: THE DEATH OF CLUBFOOT

CLUBFOOT WAS A LEGEND IN HIMSELF. HE WAS THE LAST KNOWN GRIZZLE BEAR IN ARIZONA. LIKE MANY OUTLAWS OF THE OLD WEST WHO GOT BLAMED FOR CRIMES THEY DID NOT COMMIT, CLUBFOOT GOT THE BLAME FOR ALMOST EVERY MAULED COW THAT OCCURRED IN THE WHITE MOUNTAINS. BUT, THE BEAR HAD REASON TO HATE MAN. WHEN IT WAS A YOUTH, IT WAS TRAPPED IN THE IRON JAWS OF A TRAP. TO SURVIVE, THE BEAR EITHER PULLED OR CHEWED ITS RIGHT FRONT FOOT OFF. THUS HIS NAME—CLUB FOOT. THE BEAR LEARNED TO HATE THE SENT OF MAN OR ANY THING THAT HAD TO DO WITH MAN. IT SOMEHOW LINKED CATTLE WITH MANKIND, AND STARTED A WAR OF MAULING CATTLE. CLUBFOOT WAS KILLED IN 1925 BY GENE WAITE (IN AN ACT OF SELF DEFENSE). BUT, THE MAULINGS AND MOGOLLON MONSTER SIGHTINGS DID NOT CEASE. EVEN THOUGH THIS STORY IS THAT OF A BEAR, I THOUGHT IT WAS A WORTHWHILE CONTRIBUTING STORY TO THIS BOOK.

It was a beautiful spring morning. Gene Waite prepared to take his three nephews into the mountains to check the traps they had set out the week before. The four set out into the wild area around Mount Baldy. Their trap line was along the West Fork of the Black River outside of the Indian Reservation.

"Here's another empty trap," Levi called. "It looks like it was messed up."

Gene shook his head then replied in despair. "That dang bear!"

"Do you think we ought to go back and get our big guns?" Michael questioned Gene.

"No I don't think so," he replied. "The bear is probably long gone."

Gene picked up his twenty-two single-shot rifle. He always carried this gun for small game. It was a good way to pick up some extra meat while he was out checking the traps. It was just right for the opportune rabbit or squirrel which he may came across. The four headed down the trail to the last trap. From there, it would be a short trip over the hill to their vehicle and back home. Unfortunately, there had been nothing in the traps and no animals to shoot.

"The soup is going to be mighty watery," Gene mumbled to himself. "That dang bear has managed to find every trap along the way."

The bear had done just that. It found each trap, sprung it, and then ate any bait which was used. Afterwards, the bear had set about trying to mangle the traps. Sometimes it had succeeded.

Gene and the boys moved slowly down the trail to the next trap. This was out of necessity. Gene had a double hernia. He was not about to allow any sawbones to cut him open to fix the problem. Instead, he had elected to wear a truss to hold it in. He often commented how it looked much like his wife's girdle. The town doctor had told him not to exert himself in lifting anything heavy and no heavy hiking. This was the reason for him bringing his three nephews.

Eventually, they made the area where the last trap was located. Michael was the first to find it.

"It's still set." Michael called. "The bear hasn't been here."

"I don't like the sounds of that," Gene mumbled to Red. "Either we beat it here, or the bear is still in the area."

Red studied the area closely. His eyes told of his nervousness.

"Uncle Gene," Red insisted. "Maybe we ought to get out of here."

"Nonsense, boy." Gene commented. "That bear is probably more scared of us than we are of him."

The two moved over to a large stump and Gene sat down.

"Got a joke for ya," Gene commented.

"What is it."

"There was two hunters in the woods," Gene began. "They were out hunting rabbits when they stumbled on a great big bear. The bear charged them."

Red nervously smiled.

"One of the hunters started to run but noticed his partner was pulling off his boots.

"Why are you pulling off you boots?" the runner called.

By this time the second hunter had finished pulling off his boots and started to run. As he passed by the first, he called; "I don't have To out run the bear, I just have to out run you!"

"Here," Gene reached into his pouch and pulled out a pig's hooves and handed it to Red. "Take this over to Michael and have him re-bait the trap."

The young lad complied and was soon busy trying to help Levi and Michael reset the trap. Gene sat back and enjoyed the sun and breeze.

Moments later, Gene heard a noise on the hill above him. He turned around to see what had made the noise. He was startled to see the grizzly standing on its hind legs watching the boys.

"Boys!" Gene urgently shouted. "Run for the car!"

The boys looked up from the trap and saw the bear on the hillside. They were off and running in an instant. Gene grabbed his rifle and began to follow his nephews. However, he was not as fast as the boys. He looked back over his shoulder and saw the bear charging down the hill side. It meant business, and it was after him.

Gene tripped. He fell flat on his face. He felt his truss break. The sudden sharp pains told him his hernia had bulged, and he was not going anywhere.

Gene managed to pull himself to a log and prop himself up looking back at the oncoming bear. He took the safety off his twenty-two and waited.

Seconds later, Clubfoot came rumbling up the trail. The bear came to a halt when it spotted Gene laying in the trail. At first, it was surprised, but it didn't take long to realize it had just been given a free meal. The bear stood up on its hind legs and began to roar its victory growl.

Gene, knew he had one chance and took careful aim. The shot rang out, but was nearly drowned out by the bear's roar. The bear suddenly stopped its roar and closed it's mouth. It acted much like a human would if he would have swallowed a fly while singing an opera. It looked down at the human and collapsed.

Gene's shot was one in a million. He had managed to place the bullet through the roof of the bear's mouth into the brain. It was the only place a twenty-two shell could have fatally penetrated the bear's skull.

The bear fell forward onto Gene. The only thing keeping Gene from being crushed under the massive weight was the log which Gene had propped himself up with. He had managed to partially duck under the wood when the bear fell.

A half an hour later the three boys came back carrying their big guns. They were fully prepared to find the bear eating the remains of their uncle. Instead, they found the bear.

There was no uncle.

"Uncle Gene!" Red called. "Where are you?"

No answer came.

"He's got to be here somewhere." Levi stated. "But what killed the bear?"

The three approached the bear with great care. The bear seemed to be talking.

"Help me." a faint voice came.

"You don't suppose it ate Uncle Gene?" Levi commented.

The bear moved slightly, and the boys jumped back with their rifles ready.

"Get me out of here!" came a voice from the bear.

It was then that the boys realized, Uncle Gene was under the bear.

CLUB FOOT

CHAPTER 12: OUR STORY AND HOW WE FOUND THEM

WE DECIDED TO CHECK OUT A LOCATION IN WHICH WE HAD RECEIVED SEVERAL FRESH REPORTS. IT WAS OUR FIRST OUTING TO LOOK FOR EVIDENCE OF THE MOGOLLON MONSTER. MEMORIAL WEEKEND, 2008. WE CROSSED OVER FROM REPORTS OF SIGHTINGS AND STORIES TO ACTUAL FIELD RESEARCH.

Desert Rat (DR) and myself decided to check out a well and spring that is about ¼ mile above the Falls. We worked the area looking for tracks and any sign of the Mogollon Monster. It was only going to be a quick over-nighter, and we were going to return on Memorial day.

As usual, we saw nothing around the spring or well. Of course, and elephant could have walked through that area and not be noticed because of all the pine needles on the ground from the very dense growth. The day was winding down and we decided to find a campsite in a small clearing. We were not sure of the fire restrictions, so we decided not to start a fire. Especially, since we were very near the Forest Service lookout tower. It was getting late and we didn't want to look for wood in the dark.

I got busy with dinner on the propane stove, while DR set up the dome tent and air matrices. It was a large 6 man tent with the flexible rods running up over the top to create the dome. The rain fly was installed over the top when the dome art was up and the tent was staked out. All-in-all, a pretty stable tent.

After the meal and cleanup of dinner, we decided to turn in early. I decided to for-go the usually ritual of placing my motion detectors, game camera, and voice activated tape recorder on the edge of camp. In fact, I left them locked in the truck along with my hand-held camera. But, I was tired and figured it was safe because we had combed the area pretty well during the day. Nothing was out there. I'm supposed to be the expert (definition of an expert: A drip under pressure).

About 2:30am I awoke to a very foul stench. It smelled like decomposing fish. It was very strong and was turning my stomach. I was laying on my back facing up when I opened my eyes. Instantly, I noticed the tent side was about two inches from my nose. I was terrified. Something was actually pushing my side of the tent down on top of me.

I slid my hand out of the bedroll and located my pistol while quietly calling DR to wake up. Suddenly, the tent flipped back up into the dome position. Of course, I screamed my lungs out, and rolled to the center of the tent.

I never saw a man move so fast as DR did. He was out of the bed and had pistol in hand almost before I reached the center of the tent. He started making very angry noises and hitting the sides of the tent with his pillow. I calmed down enough to turn on an electric lantern, and threw it outside. This allowed us to see under the rain fly a short distance outside, but hopefully nothing outside could see our shadows from the inside.

Nothing was near the tent. I must admit, if anything would have touched the tent at that point, we would have blasted it, and the tent would have had a new door!

We waited for a few moments (which seemed like an eternity to me), then DR slipped out side. He picked up the lantern and did a quick walk around.

Me? I headed straight for the truck. Once inside, I started the engine and turned on the lights. I was ready to hit the gas if anything was spotted, all the while having the door open for DR to jump in.

We decided to go to the nearest town and come back in the daylight to get our tent and stuff. We found no foot prints, but there were some places where the pine needles had been pushed back on my side of the tent.

I still quake in my boots whenever I think about it. However, I do kick myself for not setting my motion detectors, game cameras, and recorder. And, if we would have seen the creature, I would not have taken any pictures. That was the last thing on my mind. I would have been too busy beating a hasty retreat! I also know that I sound just about like the woman in the horror film when they get surprised by the monster.

I wonder why my side? Was it curious? Or Hungry?

A week later an expedition of four (DR, Preston Smith, and Alex Hearn) returned to the tent site to investigate the event. While searching the area, they found many impressions that seemed to be made by extremely heavy and large feet. A couple of bedding areas were located where branches had been piled up under the low brows of a tree to make a bed. There were broken branches with leaves laying all around the area.

As the group made their way back to the road, they ran into a Forest Ranger, and they stopped to talk for a while.

The Forest Ranger (Dave) stated there had been no bear in the area for two years, but he had been finding scat that he could not identify. He had not thought much about it until he ran into us. While we were talking to the Ranger, Kyle located our first major find.

As the Ranger drove away, Kyle called us over to the side of the road. We were dumb stuck with what he had found. It was a footprint--19 inches long, 9 inches through the ball and 4 inches wide in the heel. Alex provided the plaster to cast the print.

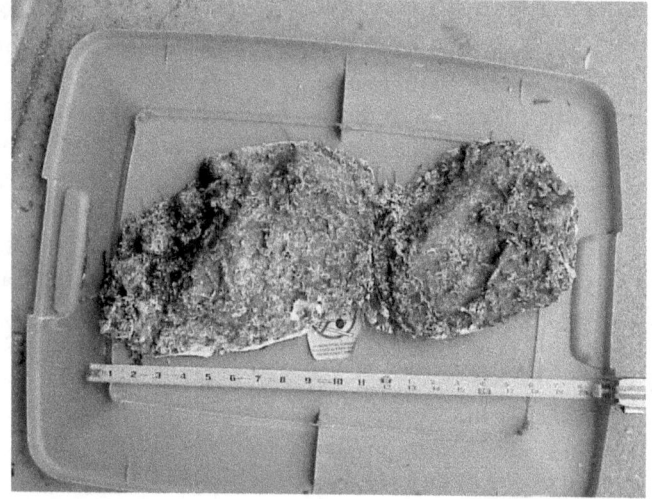

We determined that this area needed further investigation.

EXPEDITION 1
6-7 June, 2008 (Friday-Saturday)
Participants: Kyle Barentine, Preston Smith, Alex Hearn, Mitchell Waite
Purpose of the trip: Two weeks earlier, Mitch and Susan went camping in the Mogollon Rim area. During the middle of the night, the dome tent was squashed down on Susan's side and held about two inches above her nose. The

tent was released and popped back up to the dome shaped. During this time there was a very foul odor which Susan believes was that of the Mogollon Monster (Arizona's Bigfoot). Both made an evacuation of the tent and made it to their vehicle, but saw no animals. They left the area to return during the day to pack up their belongings. Two weeks later the expedition was organized to go back and try to find out what had happened.

Departure and arrival at camp: Preston, Kyle, and Mitch Waite departed Mesa, Arizona at approximately 4:25pm on Friday, 6 June, 2008.

We arrived at the Base Camp at approximately 6pm and went about setting up camp. Once the tent was up, Kyle and Preston went exploring the area.

Kyle and Preston find Turkey Bones: Up the trail from Campsite 1 they located the bones and feathers of a Turkey. The odd thing was the bones were picked clean much like when someone sits down and eats a Thanksgiving dinner. A bear, cat, or any other predator, will beat bones and all. We decided to set up our first game camera on these remains hoping whatever would return and we would get a picture of the creature. Later we find two more turkey kills. Again, bones were all intact and not chewed.

Mean while I was working on getting my equipment working. Night vision scopes and motion sensors needed batteries, Game cameras needed SD cards, etc. A final check and they were ready to go.

I walked up to where the turkey remains were located and set the camera looking up the small game trail leading to the camp.

I set up the motion sensors so we could tell if anything was coming down the trail into our camp. Unfortunately the only 9 volt battery I had available was too week to operate the sensor with any distance. It was virtually useless. We would have to do without the early warning it would give if our turkey eating visitor decided to come into our camp. I came back to camp and put two propane lanterns hanging from tree limbs to light up the campsite for the night. They were positioned as to leave no dark zones around the tent and cooking area.

As I finished placing the last propane lantern, Kyle and Preston returned to camp and started working on dinner. That is when the first rock came down from the sky just missing the hood of my truck. It landed with a very definite thud, and was about the size of an golf ball. It was not totally dark, but we were losing light so I fired up the lanterns.

Just before dark we heard Alex Hearn squawking on the walkie-talkies. He was in his car headed for our camp and trying to reach us to see if we were already in camp. A short time later he arrived at the campground. When Alex arrived we showed him around the area and went to the first turkey kill to show him our camera set up. Afterwards, we returned to camp for dinner. It was good and tasty.

After dinner, Alex started setting up his truck for sleeping. He was going to spend the night in his tuck. As he was headed for the vehicle a second rock came crashing through the trees landing just short of him. It rolled to a stop. The rock came from uphill.

When the rock fell, we all froze. Instinctively we were listening for any clues of an attack. Movement on the hill, told us we were not alone. We were being watched. Of course we had a few tricks up our sleeve. We broke out the night vision scopes and began to study the hill side above us. Slowly, we began to venture away from the circle of light around our camp. We found nothing.

Being the first time we had ever encountered these circumstances we retreated back to camp when we didn't see anything. We were standing around the fire when the first vocalization was heard. It sounded like a whoop type call. It was not too far to the west of our camp. It sounded several times and each person in the camp got to hear it. This was not a pack of coyotes or wolves. It definitely was not an elk. We all agreed on this.

The vocalizations ceased, and we were talking about what we had heard. Alex picked up a large solid stick and took a whack at a tree. A few moments later, he did it again but this time three quick hits.

We listened for a few moments and suddenly the vocalizations retuned. This time there multiple calls. It sounded like two to the west of us coming our way. Then another group sounded from the South. We were getting quite excited about this when the third group chimed in from the North east. We were surrounded. The calls lasted for several minutes and ceased. Things in camp began to return to normal. We had just had some very interesting events and our senses were pretty much on over load. Eventually, we one by one dropped out and went to bed.

About an hour before day break we (in the tent) awakened to a strange whistle. It was like nothing we had heard before. It sounded as if someone was whistling bird calls. We lay in our beds whispering to each other trying to figure out what would be making that kind of a sound. Eventually, the whistling moved around our tent and faded in the distance.

Day break came none too soon. It was good to be able to see the things around us. Breakfast came and went and we were off to check our game camera on the turkey kill. There were no pictures. The camera had not been tripped. But, the camera was working perfectly as evidenced by the picture of Kyle and Preston checking the counter on the face of the camera.

Our objective for the day was to go to where Susan had the tent collapsed down on her. She was sure it was not a bear. Her comments were, "Bears don't have hands, and they don't smell like that." We headed up the road to the area above the water falls. On our way up we noticed a mine shaft on the side of the road. It was an old Uranium mine, and still had the radiation hazard sign posted at the entrance. Of course, none of us had any inclination to see if we would glow.

We finally reached the campsite. The first thing we noticed was how dense the growth was in the area. Ferns were up to our waist. The trees were very thick. We found a small trail going down the hill side to a creek. We began to notice the amounts of broken limbs with leaves laying on the ground. It was almost as if someone had been harvesting the smaller branches with leaves. We found several piles of these limbs under tree limb shelters. The shelters were smaller trees pulled over and wedged under a much larger tree limb creating a lean-to shelter. The limbs with leaves were all stacked under the lean-to with the broken (trunks side of branch) facing up hill creating a bed.

Looking around we began to find some partial foot prints. Nothing worth casting, but there was evidence of a lot of activity in the area. We worked our way up stream on the creek and eventually came out on the road. But, just before we got to the road we found a rain gage that had been totally bent to pieces. It looked as if someone had tried to take it apart with a very big hammer.

We topped out on the road and began to head back to our vehicles. We heard a truck coming, and sure enough it was the local Forest Ranger. He stopped to talk to us for a few minutes because he was curious about my truck. My truck had some magnetic signs on it displaying pictures of the Mogollon Monster and some big feet in the windows. We explained to him why we were there and what we had been doing. He just smiled at us and said he had seen nothing strange, but had seen some scat that he had not been able to identify. He went on to say that there was a big brown bear in the area, but no one had seen it in about two years. In fact, no one had sighted any bears at all. However, there was a black panther ranging on a mountain to the south of us. So he cautioned us to be careful. He departed down the road.

As we were talking, Kyle was looking at something he had found o the side of the road not 20 yards away from where Susan had previously had her tent squashed. It was a huge footprint. We couldn't believe what we had found. We counted the toes and discussed casting the print. None of us had any experience, but Alex had come prepared with footprint casting materials. We took our first footprint casting.

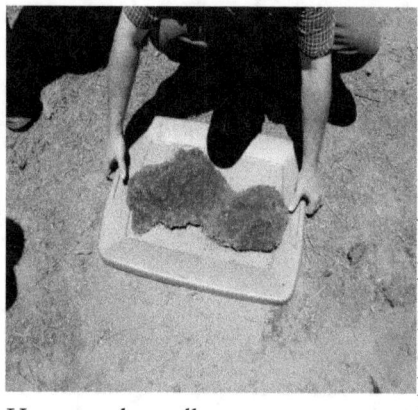

When we returned to our trucks, I was in for a surprise. A flat tire. I was very lucky to have a spare tire that was pumped up and ready for service.

We headed back home. On the way down the hill, Ranger Dave pulled us over. He wanted to talk to us some more about our sighting—not knowing about our latest find. He told us more about the scat and where it had been found. The location was not too far from where we had recovered the footprint.

Ranger Dave went on to tell us about Crazy Larry, a hermit that lived in the local caves in the area. He wanted us to be careful not to mistake him for a bigfoot. We kind of laughed and showed him the footprint. We asked him if the hermit could leave a footprint like what we had. Dave got excited. He had never seen a print like this one. He was totally amazed. He told us that there was many stories of the area about lost Indians, a jaguar, but no bigfoot. He was now a believer. Eventually we had to say good bye, and we headed for home.

We cleaned the print and got to see our success. All we had to do was spray off the plaster with the garden hose and this is what we come up with. It is not an overstep bear print. You can clearly see the toes are in a slant. Bear toes form a V with the middle toe being the leading toe. The other bear toes slant away from the middle toe. Besides, there are no bears in Arizona with a 11 inch paw (compensating for the foot ending at the arch) being 9 inches wide. Arizona has no grizzlies, no Kodiak , or no polar bears.

THE SEARCH WAS ON!